T0226111

# Advanced Technologies and Societal Change

This series covers monographs, both authored and edited, conference proceedings and novel engineering literature related to technology enabled solutions in the area of Humanitarian and Philanthropic empowerment. The series includes sustainable humanitarian research outcomes, engineering innovations, material related to sustainable and lasting impact on health related challenges, technology enabled solutions to fight disasters, improve quality of life and underserved community solutions broadly. Impactful solutions fit to be scaled, research socially fit to be adopted and focused communities with rehabilitation related technological outcomes get a place in this series. The series also publishes proceedings from reputed engineering and technology conferences related to solar, water, electricity, green energy, social technological implications and agricultural solutions apart from humanitarian technology and human centric community based solutions.

*Major areas of submission/contribution into this series include, but not limited to:* Humanitarian solutions enabled by green technologies, medical technology, photonics technology, artificial intelligence and machine learning approaches, IOT based solutions, smart manufacturing solutions, smart industrial electronics, smart hospitals, robotics enabled engineering solutions, spectroscopy based solutions and sensor technology, smart villages, smart agriculture, any other technology fulfilling Humanitarian cause and low cost solutions to improve quality of life.

More information about this series at https://link.springer.com/bookseries/10038

Ch. Satyanarayana · Xiao-Zhi Gao ·
Choo-Yee Ting · Naresh Babu Muppalaneni
Editors

# Machine Learning and Internet of Things for Societal Issues

Springer

*Editors*
Ch. Satyanarayana
Department of Computer Science
and Engineering
Jawaharlal Nehru Technological University
Kakinada
Kakinanda, Andhra Pradesh, India

Choo-Yee Ting
Faculty of Computing and Informatics
Multimedia University
Cyberjaya, Malaysia

Xiao-Zhi Gao
School of Computing
University of Eastern Finland
Kuopio, Finland

Naresh Babu Muppalaneni
Department of Computer Science
and Engineering
National Institute of Technology Silchar
Silchar, Assam, India

ISSN 2191-6853    ISSN 2191-6861  (electronic)
Advanced Technologies and Societal Change
ISBN 978-981-16-5092-5    ISBN 978-981-16-5090-1  (eBook)
https://doi.org/10.1007/978-981-16-5090-1

This Springer imprint is published by the registered company Springer Nature Singapore Pte Ltd.
The registered company address is: 152 Beach Road, #21-01/04 Gateway East, Singapore 189721,
Singapore

# Preface

This book aims to provide theoretical and practical machine learning concepts (ML) and IoT techniques for various intelligent applications. The main objective of machine learning is to identify the pattern in the present data and make predictions on future data. Huge data is generated from various sensors and devices connected through Internet, called Internet of things. Technology is bringing changes in day-to-day activities. Intelligent systems are playing a key role for the betterment of society. We have also witnessed that many smart devices are interconnected. The data generated is being analyzed and processed with machine learning models for prediction, classification, etc., to solve human needs in various sectors like health, road safety, agriculture, and education. We can turn on air conditioner at home from a remote location, our refrigerator can send alert messages to fill it with vegetables, fruits, etc., farmers can turn on the motor by sitting at home, and electricity bills are generated automatically every month like this there are much new innovation and intelligent solutions for making our life easy. This book covers a variety of machine learning, IoT applications in the fields of agriculture, health care, education, etc.

Kakinanda, India        Dr. Ch. Satyanarayana  
Kuopio, Finland        Dr. Xiao-Zhi Gao  
Cyberjaya, Malaysia        Dr. Choo-Yee Ting  
Silchar, India        Dr. Naresh Babu Muppalaneni

# Introduction

Just like electric energy, machine learning and Internet of things (IoT) will inspire our life in many ways some of which are not even susceptible today. This book provides a thorough abstract perceptive of machine learning and IoT techniques and algorithmic program, many are mathematical concepts that explained in a spontaneous manner. The book starts with a summary of machine learning and IoT and supports mathematical and statistical concepts before moving on to machine learning and IoT topics. It step by step builds up the depth, concealing many of the present-day machine learning and IoT algorithms, ending in deep learning and employ reinforcement learning algorithms. This book also covers some of the mostly used machine learning and IoT applications. The crucial part of the book in is individual to any specific programming languages or hardware so that reader can try these concepts on whichever platforms they are already familiar.

## Machine Learning

Humans are able categorize the things like books, pen, pencil, vehicles, etc., as the brain trained itself. Similarly, to make machine to understand, we need to train using the past data to predict future. Machine learning is an area of computing machine that is concerned with developing algorithmic rule along with method for providing intuitive solutions for complex problems that are otherwise hard to deal with using traditional programming methods. These traditional programming methods comprise of two defined phases. The first phase creates a detailed design for the program to be employed. It thus comprises of a fixed set of steps or rules to be followed for dealing with the problem. The later deals with implementing this design on any application in the form of a computer program.

ML solutions are challenged for many real-life application problems for creating a elaborated design that can be usually tough to deal with a given problem. Among many problems that ML needs to deal with, detecting or recognizing handwritten characters in an image one of them. Machine Learning is solving many issues in

helping humans, and one such application is the recognition of handwritten characters. By learning the elaborated design from a set of large labeled data set, the ML
can deal with such problems. The more they extract features, results become more
accurate. The goal of a machine learning algorithm is to train a model from a labeled
data set so that it can predict for unknown data points correctly.

Machine learning is a subset of artificial intelligence (AI). ML generally extracts
relevant features from the data and equips these features to the models that can be
learned, understood, and used for socialite issues. Many technologies such as facial
recognition systems, OCR, and recommendation engines have tremendous benefits
from ML.

Many social media platforms, with the utility of facial recognizers, can automatically identify faces on photos and automatically suggests tags to them. The
optical character recognition (OCR) technologies are also recognition engines that
can convert a piece of text in an image into portable format. Recommendation
engines, also a widely used application of ML, suggest users movies or shows by
learning their interests based on their previous playlists, thus while working with the
machine learning models

Applications employing ML have been highly developing and improving day by
day. Hence, one should take care of a number of things while working with these
machine learning methodologies.

## Machine Learning Methods

The machine learning methodologies are classified into two categories which are
based on learning and feedback provided to the system in order to train the model.

These two learning methods are: **Supervised and unsupervised learning**. The
former trains the model based on labelled data, while the later trains the model by
learning structures of the unlabeled data.

## Supervised Learning

For supervised learning, we need labeled data which means each record contains
attributes called features, and the label is called class. Usually, the data is divided
into training set and testing set. Employing the various machine learning algorithms
such as support vector machine (SVM), decision tree, and Naive Bayes, the training
is performed on the training dataset which means understanding the patterns using
the algorithm with minimal error with target class. In supervised learning, the model
is trained with inputs that are labeled with their corresponding outputs or labels.
The advantage of using such a model is that the model trains and rebuilds itself by
computing an error and trying to reduce the error at every iteration. It computes the
error by comparing the actual output with the predicted output. Supervised learning

tries to label or classify the unseen or unlabeled data points by learning and training on the labeled and seen data. Hence, one can say that this type of learning trains model on past data and predicts upon the future data. It can use the knowledge from the past stalk information and gain insight into upcoming fluctuations or tag unlabeled photos or images based on previously seen images of labeled ones.

## Unsupervised Learning

In contrast to supervised learning, unsupervised machine learning deals with unlabeled or unclassified data or input to train the model. So, it learns similarity patterns among the input data on its own and classifies the unseen and unlabeled data. Thus, this type of learning discovers hidden or unseen patterns in the data and then extracts relevant features from them and uses this knowledge of the features to classify new raw data.

**Unsupervised learning** is employed on basically online available data. The purchase data of items of all the clients may be available at a system, but there is no available similarity measure that relates the client's profile to the type of purchases that they make. So, this data can be given into a supervised learning model or algorithm which can learn similar patterns from the provided input data and learn relationships among the type of client and their purchases. It is supposed observed that women of a particular age group who buys unscented soaps are pregnant, and hence, a campaign relating to pregnancy and baby care products can be targeted in order to increase the purchases of the shop/market.

## Internet of Things

What if weather conditions are sensed by an umbrella and remind the user to carry it day along, or wearable devices monitoring a patient's health condition and convey the same to the doctor, or if a car could predict the system which reminds the owner about the servicing schedules to avoid malfunction before head.

These scenarios can be practically implemented in the real world with the help of **Internet of things (IoT)** and Internet connected with cloud platform. It can thus be viewed as a dynamic network having all the physical and virtual objects interconnected with each other. The main streams of IoT include cloud computing, artificial intelligence, and wireless sensor networks can be practically incorporated to healthcare, transportation, and logistics as well.

## Organization of the Book

In this book, the author walks through all the major design and implementation details of various functionalities done with IoT. It is an idea which enables communication between interworking devices and applications, where physical things communicate over the Internet. Reality is made up of smart cities and is also expected to make self-driving cars.

However despite these efforts, some issues are still challenges, such as IoT services should be increased, functions like scalability, accessing control, storage, fault tolerance, and privacy.

Main challenges still facing the Internet of things

- IoT security.
- Lack of regulation followed in IoT.
- Challenges and compatibility.
- Bandwidth is limit.
- Customer prospect.

Chapter 1 describes generator adversarial networks (GANS) which is one kind of deep learning model based on Min-Max algorithm. It is a modeled network; it translates data from images which can be utilized for image-to-image translations, semantic image-to-photo translations and text-to-image translations. They are different types of GANS, and a comparative report of various generative adversarial networks is given in terms of different Inception Scores.

Chapter 2 highlights the vast opportunity of machine learning that create in educational environment, where machine learning are deployed along with other technologies to provide fast, efficient, and quality learning experiences and also highlighted its issues and challenges to be faced by the educational environment in terms of expenditure, evaluating student's performance process and providing e-resources for online learners and challenging different approaches to dominate the digital world.

Chapter 3 describes the sensor technology for continuous monitoring of the glucose in diabetics patients using IoT. This monitoring method includes processing of data with timely readings to reduce the diabetic complications at about 35–70%. The monitored values of glucose levels for a single patient that are diagnosed are found to be accurate at all the intervals of time to overcome a stay in the hospitals.

Chapter 4 describes one of the challenging roles of machine learning and IoT in agriculture. Good farming is based on smart farming cycle which includes observation, diagnostics, decisions, and action. IoT-cloud platform collects the data from IoT devices for observations, and actions are performed. Its main application is to target at traditional farming operations and its challenges in order to satisfy rising demand and reduce production losses.

Chapter 5 describes the technologies like block chain, machine learning and IoT and proposed a model to counter COVID-19 and future sustainability. It considers the effects and severity of pandemic and provided a model to safeguard the airport which forms the central and critical point of disease spread. Maintenance is also a part of challenge in terms of health and hygiene which is monitored through IoT devices by setting up a threshold value.

Chapter 6 discusses the comparative study made with others on COVID-19 symptoms by using modified conventional neural networks. Evaluating different acoustic features of cough, breath, and speech voices were performed. The research elegantly compares and concludes a patients' voice inconvenient accuracy and is found to be proportional to his/her cough and breathe sounds. It also surfaces the main reason behind these inefficient preliminary results, as time constraints and computing power. It reduces computational cost, by working with copious amount of training data which is better than the existing system for prediction of COVID-19.

Chapter 7 describes a cyber-physical system (CPS) model, which takes care of issues occurred at toll booths. This model will help to overcome the waiting time of vehicles at tolling booths by removing all sort of human interaction. Five cases are identified and considered and reported how the system works under this cases which are identified smoothly. When electronic toll collection systems go down, human interaction is needed to handle such situations to overcome the waiting time and fuel.

Chapter 8 discussed the sentimental analysis of machine learning, where Telugu code mixed tweet (STCMT) is used extract the sentiments, and pre-processing steps are implemented; language identification of each word in the tweet is transliterated into Telugu script by using Telugu SentiWordNet, sentiments are extracted from the Telugu words, and comparative study is discussed.

Chapter 9 describes integrated neural network design for image processing and a novel approach for object recognition with visual information by considering the input shape in complex plane and represented with the measure of vortex flow on the cylinder with angle of attack. The Fourier transform is used to describe the input object, and performance measures indicate the efficiency of the proposed approach for classifying the input medical images.

Chapter 10 explains the hybrid approach using fuzzy C-means (FCM) and gradient vector flow (GVF) active contour model. It is applied to segment the tumor from the MR image of the brain. Results are collected conducting the experiment with the MRI datasets from brain web database. Performance is evaluated by using various metrics.

Chapter 11 discusses deep learning-based image segmentation which is a strong instrument in picture division. Tumors are of various structures and have various highlights and various medicines. A model is proposed to identify brain tumors using deep learning techniques. This model is built using YOLOv3 architecture, and results are impressive and have better performance when compared with existing methods.

Chapter 12 describes the effectively simulated CdS thin-film solar cell and its properties using ANN by predicting various parameters. A three-layered ANN architecture is implemented with various experimental datasets. Predictions of the parameters are carried out with various hidden layer neurons and comparative results are presented.

Chapter 13 describes the deduplication concept in cloud storage for data collected through sensor. A deduplication approach is not to have repeated sensor data. Subsequently, the sensor data will be less vital blocks in order to support the architecture. Tests are performed on 01, 05, and 10 deduplicators of planned model, and results are comparatively shocking.

# Contents

# About the Editors

**Dr. Ch. Satyanarayana** is currently working as Professor in the Department of Computer Science and Engineering, Jawaharlal Nehru Technological University Kakinada, Andhra Pradesh, India. He received his Ph.D. in 2007 from JNTU. He has 20+ years of teaching, research experience, and ten years of administrative experience in various capacities like controller of examinations, head of the department, director academic and planning. He has supervised 23 Ph.D. students and 100+ master's students. He is Senior Member of IEEE. His research interests include image processing, speech recognition, pattern recognition, network security and big data analytics and computational intelligence.

**Dr. Xiao-Zhi Gao** received his B.Sc. and M.Sc. degrees from the Harbin Institute of Technology, China, in 1993 and 1996, respectively. He obtained his D.Sc. (Tech.) degree from the Helsinki University of Technology (now Aalto University), Finland, in 1999. In January 2004, he was appointed as Docent (Adjunct Professor) at the same university. He is now working as a professor of data science at the University of Eastern Finland, Finland. He is also Guest Professor at the Harbin Institute of Technology, Beijing Normal University, and Shanghai Maritime University, China. Prof. Gao has published more than 400 technical papers on refereed journals and international conferences, and his current Google Scholar H-index is 32. His research interests are nature-inspired computing methods (e.g., neural networks, fuzzy logic, evolutionary computing, swarm intelligence, and artificial immune systems) with their applications in optimization, data mining, machine learning, control, signal processing, and industrial electronics.

**Dr. Choo-Yee Ting** is currently working as Professor at the Faculty of Computing and Informatics, Multimedia University, Cyberjaya, Malaysia. He is also Deputy Dean of the Institute for Postgraduate Studies, Multimedia University. In year 2002, Choo-Yee Ting was awarded the Fellow of Microsoft Research by Microsoft Research Asia, Beijing, China. He has been active in research projects related to predictive analytics and big data. Most of the projects were funded by MOE, MOSTI, Telekom Malaysia, MDeC, and industries. In year 2014, he and his team members won two national level big data analytics competitions. Dr. Ting has been the trainer for MDeC and INTAN for courses related to big data and data science. He is also the consultant, mentor, and assessor for projects under MDeC funding. Currently, he is working on Trouble Ticket Resolution prediction for Telekom Malaysia, AirAsia seat capacity optimization, and dengue outbreak prediction for the government of Philippines. Dr. Ting is certified in Microsoft Technology Associate (Database), IBM DB2 CDA, and the Coursera Data Science Specialization (John Hopkins University).

**Dr. Naresh Babu Muppalaneni** working as Assistant Professor in the Department of Computer Science and Engineering at National Institute of Technology Silchar. He received his M.Tech. from Andhra University and Ph.D. from Acharya Nagarjuna University. He has published more than 30 papers in different International journals, book chapters, conference proceedings, and edited research volumes. He has published five volumes in Springer Briefs in Forensic and Medical Bioinformatics. He is a fellow of IETE, life member of CSI, member of ISCA, and senior member of IEEE. He is a recipient of Best Teacher Award from JNTU Kakinada. He has completed research projects worth of 2 crore rupees from DST, DRDO. He has organized six international conferences and four workshops. His research interests are artificial intelligence in biomedical engineering, human and machine interaction and applications of intelligent system techniques, social network analysis computational systems biology, bioinformatics, and cryptography.

# Chapter 1
# A Review on Generative Adversarial Networks

**Shyamapada Mukherjee, Ayush Agarwala, and Rishab Agarwala**

## Introduction

Generative adversarial network (GAN) has become an interesting topic in the field of research. As the use of GAN has increased in various domains like computer vision, time series synthesis, semantic segmentation, and natural language processing, many papers on different domains of GAN are published according to Google Scholars. The basic structure of a GAN consists of two models: (1) generator and (2) discriminator. Normally, neural networks are used to implement both the above-mentioned models, but any type of differentiable system that maps data from one space to the other can be used for making these models. The GAN uses a mini-max algorithm as the work of the generator is to create new data capturing the distribution of true dataset while the work of discriminator is to compare whether the new data generated is accurate to the original dataset, and it stops when it reaches Nash equilibrium. GAN has shown excellent improvement in the field of computer vision making complex tasks such as image generation, image-to-image translation, image completion, and image super-resolution. The difficulties that GAN faces are as follows:

1. Difficulty in training—The generator and discriminator are unlikely to reach Nash equilibrium during the training period.
2. Difficulty in evaluation—The differences between generated distribution and real distribution are considered a method for evaluation of GANs, and we are lacking in good evaluation metrics for accurate estimation.

Through this paper, our aim is to give a detailed description of GAN and the different types of GANs used for text-to-image translation. We also compared the various types of GANS on the basis of their accuracy, performance on various generalized datasets. In the first section, we have provided a detailed description of basic GAN; in the second section, we provided the description of various types of GAN for their application in different fields. In the third section, we summarized the problems and

S. Mukherjee (✉) · A. Agarwala · R. Agarwala
National Institute of Technology Silchar, Silchar, Cachar, Assam 788010, India
e-mail: shyama@cse.nits.ac.in

© The Author(s), under exclusive license to Springer Nature Singapore Pte Ltd. 2022
Ch. Satyanarayana et al. (eds.), *Machine Learning and Internet of Things for Societal Issues*, Advanced Technologies and Societal Change,
https://doi.org/10.1007/978-981-16-5090-1_1

benefits of GAN models for image generation, and at last, future aspects of GANs are discussed.

## Basics of GAN

Generative adversarial networks or GANs are one of the most dynamic zones in profound learning innovative work because of their unfathomable capacity to create synthetic results. Generative adversarial networks (GANs) are an energizing ongoing advancement in AI. GANs are generative models: They make new information occasions that take after the preparation information. For instance, GANs can make pictures that appear as though photos of human countenances, despite the fact that the appearances do not have a place with any genuine person. GAN did not imagine generative models, yet rather gave a fascinating and advantageous approach to learn them. They are classified as "ill-disposed" in light of the fact that the issue is organized with the end goal that two substances are going up against each other, and both of those elements are AI models. A generative adversarial network (GAN) has two sections as follows:

1. The generator figures out how to create conceivable information. The produced occurrences become negative preparing models for the discriminator.
2. The discriminator figures out how to recognize the generator's phony information from genuine information. The discriminator punishes the generator for delivering doubtful outcomes.

When preparing starts, the generator delivers clearly counterfeit information, and the discriminator rapidly figures out how to tell that it is phony.

### *Training Data for Discriminator*

The training data for discriminator are collected from the following sources:

1. Genuine information occurrences, for example, genuine pictures of individuals. The discriminator utilizes these cases as positive models during preparing.
2. Counterfeit information occasions created by the generator. These are utilized by the discriminator as negative models during preparing.

### *Training the Discriminator*

Two loss functions are connected by the discriminator. The loss incurred by the generator is ignored during the training of discriminator. The discriminator only uses its own loss. During the training, genuine and phony, both the information from the generator is characterized by the discriminator. There is a possibility that the discriminator misclassifies a genuine case as phony or a phony occasion as genuine. The discriminator refreshes its loads through backpropagation from the discriminator misfortune through the discriminator arrange.

## *The Generator*

The generator part of a GAN figures out how to make counterfeit information by fusing criticism from the discriminator. It figures out how to cause the discriminator to order its yield as genuine. Generator preparation requires more tightly coordination between the generator and the discriminator than discriminator preparation requires. The part of the GAN that prepares the generator incorporates the following:

1. Irregular information.
2. Generator arrange, which changes the arbitrary contribution to an information occurrence.
3. Discriminator arrange, which orders the created information.
4. Discriminator yield.
5. Generator misfortune, which punishes the generator for neglecting to trick the discriminator.

## *Using the Discriminator to Train the Generator*

To prepare a neural net, we modify the net's loads to diminish the blunder or loss of its yield. In our GAN, in any case, the generator is not legitimately associated with the misfortune that we are attempting to influence. The generator takes care of into the discriminator net, and the discriminator creates the yield we are attempting to influence. The generator misfortune punishes the generator for creating an example that the discriminator organizes arranges as phony. This additional lump of the system must be remembered for backpropagation. Backpropagation modifies each weight the correct way by computing the weight's effect on the yield—how the yield would change on the off chance that you changed the weight. Yet, the effect of a generator weight relies upon the effect of the discriminator loads it takes care of. So backpropagation begins at the yield and streams back through the discriminator into the generator. Simultaneously, we do not need the discriminator to change during generator preparation. Attempting to hit a moving objective would make a difficult issue much harder for the generator. So the generator is trained with the accompanying system as follows:

1. Test irregular clamor.
2. Produce generator yield from the inspected irregular commotion.

3. Get discriminator "genuine" or "counterfeit" grouping for generator yield.
4. Figure misfortune from discriminator arrangement.
5. To get the slopes, both discriminator and generator backpropagate.
6. Use angles to change just the generator loads.

This is one cycle of generator preparation.

## Attentional Generative Adversarial Network

Attentional generative adversarial network (AttnGAN) has permitted a multistage refinement of fine-grained text-to-image creation using attention-driven approach. The model has been divided into two parts. An attention mechanism is used in the first part of an attentional generative network. The generator uses this mechanism for constructing various parts or regions of the image by concentrating on most specific words to the sub-region being considered. It encodes the text description of images into a sentence vector which is global in nature. Each word of a sentence is converted into a word vector. Initially, a low-resolution image is constructed from the global sentence vector by the generative network. It then utilizes the image vector present for each region or sub-region for querying word vectors. This is done by an attention layer which is used for forming a word context vector. A multimodal context is formed by the combination of regional image vector and corresponding word context. Based on this, new image features are created for the surrounding sub-regions of the image by this model. This model is capable of constructing a higher-resolution image with particular details for each level. The AttnGAN has a deep attentional multimodal similarity model (DAMSM) in the other part. The DAMSM is utilized for finding the similarity existing between the image being generated and the sentence using both the global sentence level and the fine-grained word-level information. Many of the state-of-the-art GAN models have been beaten by the performance of AttnGAN model. The AttnGAN defeats other model by 14.14% increase in inception score on the CUB dataset. It also defeats other models by 170.25% inception score on COCO dataset. Here the model is taken from [1]. GAN framework provides generators and discriminators to devise the various relationships efficiently between a far apart spatial regions. The proposed model is called self-attention generative adversarial networks (SAGAN) due to the presence of self-attention nature in it. The model has been effective in modeling long-range dependencies. Spectral normalization on the generator balances the training. The TTUR accelerates the training of ordered discriminators. A performance-based comparison has been studied between SAGAN with the state-of-the-art GAN models [2, 3] on ImageNet for the generation of class conditional images. SAGAN yields the best inception score along with intra-FID and FID. The inception score goes up to 52.52 from 36.8 reported in the best-published articles. It proves its efficiency by lowering the FID up to 18.65 and intra-FID up to 83.7. This indicates that SAGAN is better in the approximating process for the distribution of original images. This achievement becomes possible and feasible

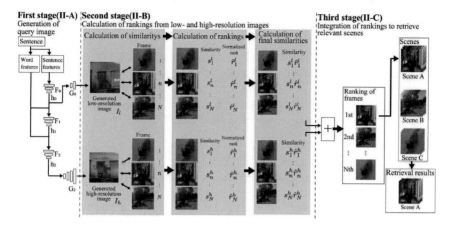

**First stage(II-A)** | **Second stage(II-B)**
Generation of query image | Calculation of rankings from low- and high-resolution images
| **Third stage(II-C)**
Integration of rankings to retrieve relevant scenes

**Fig. 1.1** Working model of control GAN

by the application of self-attention module which is able to devise the long-range dependencies between image regions. The equations helping in describing the model are taken from [4] where $\mathbf{x}$ represents image features, $\mathbf{f}$ and $\mathbf{g}$ represent feature space, $\beta_{j,i}$ gives the hold of model to $i$th position while synthesizing $j$th position, $\mathbf{C}$ denotes the number of channels, whereas $\mathbf{N}$ denotes the total number of locations of features, $o_j$ is output of attention layer, $y_i$ is final output scaled by a parameter $\gamma$ that is learnable scalar (Fig. 1.1)

$$\beta_{j,i} = \frac{\exp(s_{ij})}{\sum_{i=1}^{N} \exp(s_{ij})} \tag{1.1}$$

where $s_{i,j} = f(x_i)^r g(x_j)$.

$$o_j = v \left( \sum_{i=1}^{N} \beta_{j,i} h(x_i) \right), \tag{1.2}$$

where $h(x_i) = W_h x_i$, $v(x_i) = W_v x_i$.

$$y_i = \gamma o_i + x_i \tag{1.3}$$

$$L_D = -E_{(x,y)\sim p_{\text{data}}}[\min(0, -1 + D(x, y))]$$
$$\quad\quad - E_{z\sim p_z, y\sim p_{\text{data}}}[\min(0, -1 - D(G(z), y))] \tag{1.4}$$

$$L_G = -E_{z\sim p_z, y\sim p_{\text{data}}} D(G(z), y) \tag{1.5}$$

## Control GAN

Control GAN [5] which is the adoption of the multistage attention GAN is able to create and control the creation of images on the basis of natural language representations. The control GAN extricates various ocular features and allows parts of the artificial image to be changed perfectly, with the generation of other preserved contents. In this model, three concepts are introduced as follows:

1. The model generators which perfectly extricate various visual attributes are word-level spatial and channel-wise attention-driven generators.
2. This model introduces the word-level discriminator which provides fine-grained training signals to the generator attached to each visual attribute.
3. The randomness involved in the generation is reduced by the adoption of perceptual loss. It helps reconstructing the contents connected to the unmodified text by enforcing the generator.

A scene retrieval method [6] is used which effectively uses images generated by AttnGAN. The first step is to generate low and high-resolution images from an input sentence based on AttnGAN. The low-resolution image generated focuses on all of the input sentences while the high-resolution image generated focuses on each word of the input sentence. By the use of these different resolution images, the focus is made on the whole sentence and each word. The proposed methods notice a complex scene retrieval that can distinguish a small change between very likely scenes. Two contributions have been made in this paper. The first one is that we manifest the availability of the generated images for a scene extraction task even if the generated images are not optically nice. The second one is to make use of two generated images of different resolutions that focus on the input sentence and its words to realize highly accurate scene retrieval.

## DC-GAN

DC-GAN mainly comprises convolution layers without max pooling or fully connected layers. Convolutional stride and transposed convolution are used for downsampling and upsampling. The major architecture recommendations for working deep convolutional GANs taken from [7] are as follows:

1. Replacement of any pooling layers with fractionally stridden convolutions (generator) and strode convolutions(discriminator).
2. Both the generator and the discriminator use the batch norm.
3. The fully connected hidden layers are removed in deeper architectures.
4. Except output layer, the ReLU activation function is used in the generator. The output layer uses tanh activation function.
5. The discriminator uses Leaky ReLU activation function for all the layers.

---

**Algorithm 1** DCGAN

---

**Input:** Minibatch images x, matching text t,mismatching $\hat{t}$, number of training batch steps S

1: **Begin**
2: **for** n = 1 **to** S **do**
3:     $h = \varphi(t)$
4:     $z \sim N(0, 1)^Z$
5:     $\hat{x} = G(z, h)$
6:     $s_r = D(x, h)$
7:     $s_w = D(x, \hat{h})$
8:     $s_f = D(\hat{x}, h)$
9:     $\mathcal{L}_D = \log(s_r) + (\log(1 - s_w) + \log(1 - s_f))/2$
10:     $D = D - \alpha \mathcal{L}/\theta D$
11:     $\mathcal{L}_G = \log(s_f)$
12:     $G = G - \alpha\theta\mathcal{G}/\theta G$
13: **End For**
14: **End**

---

(DC-GAN) [8] is conditioned on text features encoded by a hybrid character-level convolutional recurrent neural network. The feed-forward inference is performed by generator network G and the discriminator network D conditioned on the text feature. It uses an understandable and productive method for text-based image synthesis using a character-level text encoder and class conditional GAN. It evolved a novel architecture and learning strategy that gave powerful visual results. It mainly focused on fine-grained image datasets. DC-GAN uses text Caltech-UCSD Birds dataset for text descriptions and a human-created caption dataset, i.e., Oxford Flowers dataset with five captions per image. It is trained and tested on class disjoint sets, and test performance gave a powerful indication of generalization ability which we also demonstrate on MS-COCO images with multiple objects and various backgrounds. Algorithm 1 shows the working of DC-GAN taken from [8].

The line (1) of the algorithm encodes the matching text description. In line (2), a sample of random noise is drawn. Line (3) forwards the sample through the generator. Lines (4) to (8) represent real image and right text; real image and wrong text; fake image and right text, respectively. The discriminator is updated in line (10). In line (12), the generator is updated.

For image generation purposes, DC-GAN showed the study about a ranking of representations from object parts to scenes. The improvement of DC-GAN is acknowledged to their three architectural contributions to CNN from [9].

1. For learning spatial down and upsampling, stridden convolutions replace deterministic spatial pooling.
2. For allowing deeper representations, fully connected layers supporting convolutional layers are removed.
3. Batch normalization, which makes the input for having zero mean and unit variance, tries to learn important functions (Fig. 1.2).

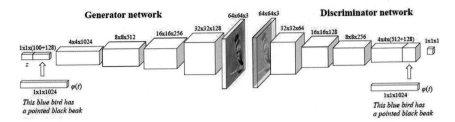

**Fig. 1.2** Architecture of the generator and discriminator networks of our DC-GAN synthesis model [9]

## Conditional GAN

Conditional GAN is a type of generative adversarial network. For a conditioned generative model, it has no authority on the types of data that are generated. For directing the data generation process, we can condition the model on some additional information. Conditioning can be on class types, on data from a different modality, or on some part of data for in a painting. Generative adversarial networks can be transformed to a conditional generative adversarial networks if both the discriminator and generator are constrained on some additional information $z$ [10], and x and y present the inputs in the discriminator. The following objective function represents a two-player mini-max game:

$$\text{Min}_G - \text{Max}_D[V(D, G)] = \mathcal{E}_{x \sim p_{\text{data}}(x)}[\log D(x|y)]$$
$$+ \mathcal{E}_{z \sim p_z(z)}[\log(1 - D(G(z|y)))] \qquad (1.6)$$

Figure 1.3 depicts the architecture of a simple conditional GAN. The method for the text-to-image problem can be answered by multiconditional GAN (MC-GAN) using a synthesis block whose work is similar to a pixel-wise gating function which controls the amount of information from the base background image by the help of the text description for a foreground object. The text-to-image problem uses a text description as an input to generate an image. It can generate an image with the attributes as per the wish of the user, as the text is used to express detailed high-level information on the appearance of an object. The raw text is embedded by a process which uses a mix of convolutional neural network (CNN) and recurrent neural network (RNN) structure. Conditional GANs are used for several image translation problems like day tonight, horse to zebra, and sketch to portrait. Here, compositional GAN is used for creating a composite image if two individual object images are given [11] proposes to solve this challenge by the following process (Fig. 1.4).

For exploring the usability of conditional generative adversarial networks [12] checks the procedure on various steps and datasets which includes vision jobs like semantic segmentation and graphics tasks like photo generation. Following points define the generality of C-GAN:

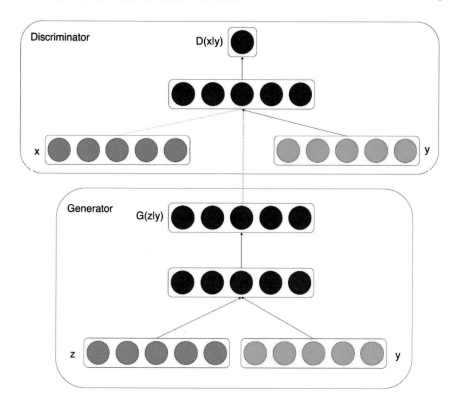

**Fig. 1.3** Simple conditional adversarial net

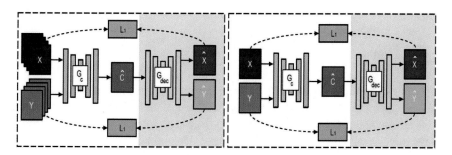

**Fig. 1.4** Generating composite image from two individual object images [11]

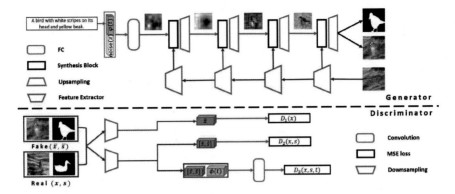

**Fig. 1.5** Basic building block of MC-GAN [16]

1. For semantic labels, it is trained on the city scapes dataset.
2. For architectural labels photos, C-GANs are trained on CMP facades.
3. For the map aerial photos, they are trained on collected data from Google Maps.
4. For black–white and colorful photos, they are trained on real images.
5. For edge photos, C-GANs are trained on edge specified image data generated using the HED edge detector plus post-processing.
6. For sketch photos, it tests edge photo models on human drawn sketches from [13].
7. For thermal color photos, they are trained on data in [14].
8. For the photos with missing pixels in painted photo, C-GANS are trained on Paris Street View [15] (Fig. 1.5).

## Cycle Consistent GAN

For solving the problem of text-to-image translation, cycle consistent GAN utilizes the working of CycleGAN. It first breaks the image synthesis network into two divisions. In the first division, a low-resolution image is generated. The generated image is provided as input to the generator in the next stage. The following division is used for refining the low-resolution input and for generating higher-quality image with 128 × 128 resolution. Cycle consistent GANs have shown brilliant results for problems involving multimodal learning. CycleGAN is a method to train unsupervised models for translating images using generative adversarial network on the basis of collected images from two different domains. CycleGAN has been applied on various applications which include style transfer, transformation of objects, season translation, and creation of images from paintings. The paper [17] addresses cross-caption cycle consistent image creation as a cascading procedure where a sequence of generators uses different captions orderly one at a time for generating images. The created image at every stage is the previous image's function and also the function of the caption provided at the recent stage which allows each stage for being constructed upon the interim images created in the just last stage, by the use of additive logic from recent captions noticed in the current level. At every level, a different CCCN and discriminator are explored. The discriminator recognizes whether the created image is counterfeit or real. The CCCN is used for translating the image to its caption and for checking its closeness to its next caption. A set of convolutional blocks makes up the architecture of the network. The first convolution layer of each block takes a caption. Each generator and CCCN diverts from the last layer of each block. A new block connects itself to expand. The number of blocks is fixed in the process of designing the architecture. A restricted number of captions are used to construct an image. The paper [18] addresses the issue of solving text-to-image problems using CycleGAN by utilizing two-stage GAN. It trains the Stage-1 GAN by maximizing LD1 and minimizing LG1 given in equations below.

$$\mathcal{L}_{D_1} = E_{(I_1, \psi(t)) \sim p_{\text{data}}}[\log D_1(I_1, \psi(t)]$$
$$+ E_{I_1 \sim G_1, \psi(t) \sim p_{\text{data}}}[\log(1 - D_1(I_1, \psi(t)))] \quad (1.7)$$
$$\mathcal{L}_{G_1} = E_{z \sim N(0,1), \psi(t) \sim p_{\text{data}}}[\log(1 - D_1(G_1(z, \psi(t))))] \quad (1.8)$$

It trains the Stage-2 GAN by maximizing LD2 and minimizing LG2 given in equations blow.

$$\mathcal{L}_{D_2} = E_{(I_2, \psi(t)) \sim p_{\text{data}}}[\log D_2(I_2, \psi(t)]$$
$$+ E_{I_2 \sim G_2, \psi(t) \sim p_{\text{data}}}[\log(1 - D_2(I_2, \psi(t)))] \quad (1.9)$$
$$\mathcal{L}_{G_2} = E_{z \sim N(0,1), \psi(t) \sim p_{\text{data}}}[\log(1 - D_2(G_2(z, \psi(t))))] \quad (1.10)$$

The results produced from CycleGAN have proved to be very impressive. The Cycle-GAN was easily able to surpass other unsupervised image translation techniques that were available during that time. In "real vs. fake" experiments, humans were unable to distinguish the synthesized image from the real one about 25% of the time.

## FM-GAN

FM-GAN and C-GAN are seen to converge with the same speed, where green and yellow lines represent the discriminative and generative loss curve of FM-GAN, red and blue lines represent the loss curve of C-GAN, and as the training process proceeds, FM-GAN has a clear process of face generation and synthesizes photo-realistic face images. The C-GAN is capable of generating only the blurry outline of faces [19]. The paper [20] proposes a model that tries to solve the limitations of attributes by training neural language models from scratch. The limitation of the model is that for finer-grained recognition, it requires a vast number of attributes. It needs those attributes which is free from the interference of natural language. The model is trained end to end so that alignment is maintained in it with the fine-grained photograph and images with category-specific contents. The model inferences are on raw text. It provides a useful annotation and retrieval mode. The model obtains positive results for image retrieval on zero-shot text-based contents. It also outweighs the state of the art based on attribute-based images for zero-shot classification on the Caltech-UCSD Birds 200-2011 dataset. In this paper [20], a variational generative adversarial network, a mix of both variational auto-encoder and a GAN, is used for modeling images with a label and latent attributes probabilistically which are then used for synthesizing images in fine-grained categories. The proposed model has two important features.They are as follows:

1. For making the GAN training more stable, it uses an asymmetric loss function.
2. For keeping the structure of generated images,an encoder network is used with pairwise matching.

The proposed method was able to create diverse and realistic samples from fine-grained categories like images of flowers, birds, and faces.

## Stack GAN

StackGAN-v1 is a two-stage generative adversarial network that is used for generating high-intent images with likeness-realistic details. The text-to-image generation method is made of two stages. Stage-I GAN generates a low-resolution image sketching from the given text description where the generated image is of basic trivial shapes and normal colors of the object. It also draws the background layout from a random noise vector. Stage-II GAN's work is to correct the mistakes from the image generated from Stage-I, and by reading the text description one more time, it completes details of the object, yielding a high-resolution photo-realistic image. The images produced by Stage-I GAN have many missing objects and can carry many shape disturbances. Few information from the description may be skipped in the beginning which is necessary for generating photo-realistic images. The Stage-II generative adversarial network is modeled based on results obtained in Stage-I for generating high-resolution images. The Stage-II GAN generates more photo-realistic details by taking into consideration the information missing during the Stage-I GAN. The paper [21] presents a comparison of StackGAN-v1 with state-of-the-art text-to-image models on various datasets like CUB, Oxford-102, and COCO to demonstrate its effectiveness. We compared StackGAN-v2 with StackGAN-v1 to show its positive side and its negative side. The inception scores and average human ranks for the proposed StackGAN models and compared methods are reported in Table 1.1.

StackGAN-v1 model shows best inception score than the all previous GAN models. It also exhibits an average human rank on all the datasets mentioned earlier. StackGAN-v1 attains 28.47% increase in inception score over GAN-INT-CLS on the CUB dataset and 20.30% betterment on Oxford-102. The improved average human rank of StackGAN-v1 depicts that this ideology is capable of generating better realistic sample images constrained on text descriptions (Fig. 1.6).

**Table 1.1** Performance comparison among StackGAN-V1, StackGAN-V2, and the state-of-the-art GANs on CLUB, Oxford, and COCO dataset in terms of inception score and human rank

| Metric | Dataset | GAN-INT-CLS | GAWWN | StackGAN-v1 | StackGAN-v2 |
|---|---|---|---|---|---|
| Inception score | CUB | 2.88 ± 0.04 | 3.62 ± 0.07 | 3.71 ± 0.04 | 4.04 ± 0.05 |
| | Oxford | 2.66 ± 0.03 | – | 3.20 ± 0.01 | – |
| | COCO | 7.88 ± 0.07 | – | 8.45 ± 0.03 | – |
| Human rank | CUB | 2.81 ± 0.03 | 1.99 ± 0.04 | 1.37 ± 0.02 | – |
| | Oxford | 1.87 ± 0.03 | | 1.13 ± 0.03 | – |
| | COCO | 1.89 ± 0.04 | – | 1.11 ± 0.03 | – |

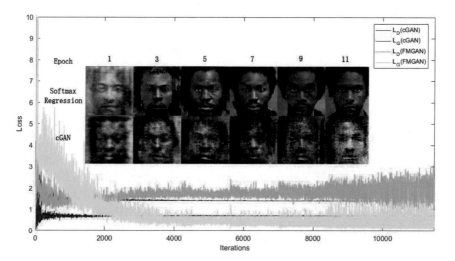

**Fig. 1.6** Generative performance comparison between FM-GAN and C-GAN during training period [19]

## MirrorGAN

MirrorGAN comprises three modules. They are STEM, GLAM, and STREAM. GLAM uses the word and sentence-level embedding generated by STEM. The cascaded architecture of GLAM creates target images from coarse-grain to fine-grain scales. It leverages word-level attention locally and sentence-level attention globally. GLAM iteratively enhances the variation and semantic balance of the generated images. The third module STREAM regenerates the text description from the created image. The generated text semantically aligns with the supplied text description. MirrorGAN explores supervised learning approach by using paired text–image data instead of training from unpaired image–image data. A mirror structure is embodied by MirrorGAN using the integration of T2I and I2T both. It uses a redescription method to utilize the concept of learning the process T2I generation. MirrorGAN redescribes an image after its generation. This process keeps the underlying semantics aligned with the original text description. MirrorGAN mainly comprises three modules called STEM, GLAM, and STREAM. As per the article, MirrorGAN achieves the best inception score on CUB and COCO datasets together. MirrorGAN Baseline yields details and consistent colors and shapes which is better in comparison with AttnGAN. In comparison of MirrorGAN with MirrorGAN Baseline, it is observable that images with more details semantic regularity are produced by GLAM. ts limitations are as follows:

1. Jointly, optimization is not possible for STREAM and other MirrorGAN modules due to unavailability of powerful computational resources.
2. A basic method is utilized for text embedding in the corresponding modules of MirrorGAN that is further improved.
3. It is believed that unlike state-of-the-art GANs, MirrorGAN is initially designed for the T2I generation by aligning cross-media semantics.

## Fusion GAN

In the paper [22], a new concept of Spatial Fusion GAN has been presented. It achieves artificial practicality in both geometry and appearance spaces. This synthesizer GAN synthesizes the geometrical shape and presentation of images. The geometrical shape synthesizer reads up the relevant geometrical of foundations of images. After that it changes and presents a closer view of objects into the background pictures collectively. The various characteristics foreground objects like color, brightness, and styles of the foreground objects are synthesized by appearance synthesizer which adjusts and implants them into foundation pictures congenially. Here a guided channel is presented for preserving the details of the image. The Spatial Fusion GANs have been evaluated based on two tasks:

1. To bring real synthesis between scene text and image pair and for training the model to turn the model a good recognition system.
2. The accuracy in scene text recognition has been evaluated over ICDAR2013 [23], ICDAR2015 [24], SVT [25], IIIT5K [26], SVTP [27], and CUTE [28] datasets. In the evaluation, one million processed text images are explored for the various procedures reported in [22] (Fig. 1.7).

Table 1.2 shows that it yields the highest recognition accuracy for most of the cases in six different datasets. It shows an improvement by 3% in terms of recognition accuracy. The results demonstrate the superiority and usefulness of its constructed synthesized images. The method depicts that the geometrical shape synthesizer and appearance processor together help to process a more rational and practical image in recognition model training. Apart from that, they are complementary, and combination of them obtains in average a 6% increase in recognition accuracy afar the baseline SFGAN(BS).

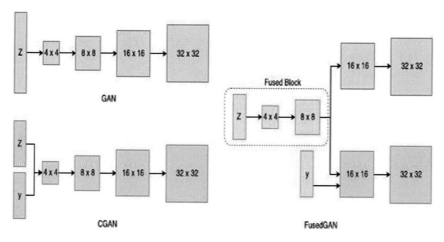

**Fig. 1.7** Detailed description of 32 × 32 image synthesis by fusing GAN and C-GAN

**Table 1.2** Scene text recognition accuracy of SFGANs has been reported over the datasets ICDAR2013, ICDAR2015, SVT, IIIT5K, SVTP, and CUTE

| Method | ICDAR 2013 | ICDAR 2015 | SVT | IIIT5K | SVTP | CUTE | AVG |
|---|---|---|---|---|---|---|---|
| ST-GAN | 57.2 | 35.3 | 63.8 | 57.37 | 43.2 | 34.1 | 48.5 |
| Spatial Fusion GAN (BS) | 54.7 | 33.8 | 63.1 | 54.3 | 41.7 | 32.8 | 46.7 |
| Spatial Fusion GAN (GS) | 56.2 | 34.5 | 65.4 | 56.6 | 42.8 | 35.0 | 48.4 |
| Spatial Fusion GAN (AS) | 57.0 | 35.3 | 65.6 | 57.4 | 44.2 | 34.6 | 49.0 |
| Spatial Fusion GAN | 60.7 | 38.1 | 68.2 | 62.2 | 47.5 | 39.5 | 52.6 |

Here the comparison is made on different types of GAN on their inception score on CUB dataset and MS-COCO dataset, and graphical representation is also given.

In the paper [29], FusedGAN is presented. It is a deep network, and it has the capability of controlling sampling of diverse images by conditional image synthesis. Good image generation model has some noted measures of quality: diversity and fidelity along with controllable sampling. High fidelity is shown by FusedGANs by performing a controllable sampling of diverse images. But there might be an argument that controllable characteristics can be shown by just losing the generation process into multiple stages. FusedGAN has an advantage over GANs. Where FusedGAN has a pipeline with a single stage along with built-in stacking of GANs, GANs multiple stages are separately trained. Existing methods requires full supervision, whereas FusedGAN produces samples with increased fidelity by leveraging abundant images without proper attention to training. Disentanglement of the generation process is

**Table 1.3** Comparison among different GANs in terms of inception score

| Types of GAN | CUB dataset | MS-COCO dataset |
| --- | --- | --- |
| GAN-INT-CLS | $2.88 \pm 0.04$ | $7.88 \pm 0.07$ |
| GAWWN | $3.62 \pm 0.07$ | – |
| StackGAN | $3.70 \pm 0.04$ | $8.45 \pm 0.03$ |
| StackGAN++ | $3.82 \pm 0.06$ | – |
| PPGN | – | $9.58 \pm 0.21$ |
| AttnGAN | $4.36 \pm 0.03$ | $25.89 \pm 0.47$ |
| MirrorGAN | $4.56 \pm 0.05$ | $26.47 \pm 0.41$ |

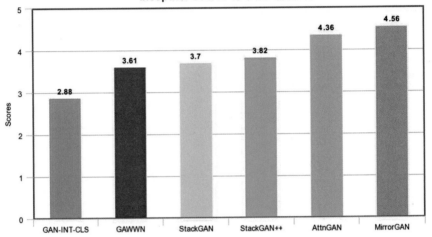

**Fig. 1.8** Bar graph presents inception score of various GANs on CLUB dataset

done, and this is achieved by the fusion of two generators performing the work of unconditional and conditional image generation. The efficiency of FusedGAN can be demonstrated by a task such as text-to-image and attribute-to-face generations.

## Comparison of Different GAN Based on Inception Score

In Table 1.3, a comparison among various generative adversarial networks has been reported in terms of different inception scores. The abbreviations "GAN-INT-CLS", "PPGN", and "GAWWN" in the table represent GAN-interpolation-matching-aware discriminators, plug and play generative networks, and generative adversarial what where networks, respectively (Figs. 1.8 and 1.9).

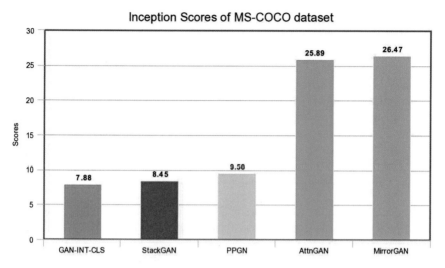

**Fig. 1.9**  Bar graph presents inception score of various GANs on MS-COCO dataset

## Merits of Generative Adversarial Network

The fundamental favorable position of utilizing generative adversarial systems (GANs) is that when it works, it works truly well, as has been appeared by the ongoing papers that produce exceptionally reasonable pictures of faces, seats, and creatures. The purpose behind this is the goal advanced by GANs—to create counterfeit information that is unclear from genuine information by another neural net—is exceptionally lined up with the objective of delivering sensible information. This is as opposed to a contending technique "variational auto-encoder" which has an apparently less-adjusted goal. Notwithstanding having a superior target, GANs do not require a ton of the earlier and back likelihood computations regularly fundamental for another contending approach, most extreme probability.

- GANs are a decent technique for preparing classifiers in a semi-regulated way. Seeing our NIPS paper and the going with code, we can simply utilize our code straightforwardly with practically no adjustment whenever you have an issue where you cannot utilize a lot of named models. Typically, this is on the grounds that we simply do not have many marked models. I additionally, as of late, utilized this code effectively for a joint effort with Google Brain on differential security for deep learning.
- GANs generate samples faster than fully visible belief nets (NADE, PixelRNN, WaveNet, etc.) because there is no need to generate different entries in the sample sequentially.
- GANs need not bother with any Monte Carlo approximations to prepare. Individuals whine about GANs being unsteady and hard to prepare, yet they are a lot simpler to prepare than Boltzmann machines, which depended on Monte Carlo

approximations to the slope of the log parcel work. Since Monte Carlo strategies do not work very well in high-dimensional spaces, Boltzmann machines have never truly scaled to sensible errands like ImageNet. GANs are in any event ready to figure out how to draw a couple of wrecked canines when prepared on ImageNet.

- Contrasted with variational auto-encoders, GANs do not present any deterministic predisposition. Variational techniques present deterministic predisposition since they advance a lower bound on the log probability as opposed to the probability itself. This appears to result in VAEs figuring out how to create hazy examples contrasted with GANs. Contrasted with nonlinear ICA (NICE, Real NVE, and so forth being the latest models), there is no prerequisite that the inactive code has a particular dimensionality or that the generator net is invertible.
- Compared to VAEs, it is easier to use discrete latent variables.
- Compared to Boltzmann machines and GSNs, generating a sample requires only one pass through the model, rather than an unknown number of iterations of a Markov chain.
- GANs are a solo learning technique: Acquiring named information is a manual procedure that takes a ton of time. GANs do not require marked information; they can be prepared to utilize unlabeled information as they get familiar with the inward portrayals of the information.
- GANs produce information: One of the best things about GANs is that they create information that is like genuine information. Along these lines, they have various uses in reality. They can create pictures, content, sound, and video that is vague from genuine information. Pictures created by GANs have applications in showcasing, Internet business, games, commercials, and numerous different enterprises.
- GANs learn thickness circulations of information: GANs get familiar with the interior portrayals of information. As referenced before, GANs can learn untidy and confounded circulations of information. This can be utilized for some AI issues.
- The prepared discriminator is a classifier: After preparing, we get a discriminator and a generator. The discriminator organize is a classifier and can be utilized to group objects

## Demerits of Generative Adversarial Network

Notwithstanding, the huge disservice is that these systems are difficult to prepare. The capacity of these systems attempt to upgrade is a misfortune work that basically has no shut structure (dissimilar to standard misfortune capacities like log misfortune or squared blunder). Hence, improving this misfortune work is extremely hard and requires a great deal of experimentation in regard to the system structure and preparing convention. Since RNNs are commonly more whimsical than CNNs, this is likely why not many (assuming any) individuals have had the option to apply GANs to much else complex than pictures, for example, content or discourse.

- Preparing a GAN requires finding a Nash harmony of a game. Once in a while slope plummet does this, occasionally it does not. We do not generally have a decent balance discovering calculation yet, so GAN preparation is insecure contrasted with VAE or PixelRNN preparing. I would contend that it despite everything feels significantly more steady than Boltzmann machine preparing practically speaking.
- It is hard to learn to generate discrete data, like text.
- Contrasted with Boltzmann machines, it is difficult to do things like an estimate of the estimation of one pixel given another pixel. GANs are truly prepared to do only a certain something, which produces all the pixels in a single shot. We can fix this by utilizing a BiGAN, which lets you surmise missing pixels utilizing Gibbs examining, equivalent to in a Boltzmann machine.
- Similarly, as with any innovation, there are a few issues related to GANs. These issues are for the most part to do with the preparation procedure and incorporate mode breakdown, inward covariate moves, and disappearing angles. How about we take a gander at these in more detail.
- Mode collapse is an issue that alludes to a circumstance where the generator organizes and produces tests that have a little assortment or when a model beginnings are creating similar pictures. Some of the time, a likelihood circulation is multimodal and complex in nature. This implies it may contain information from various perceptions and that it may have numerous tops for various sub-diagrams of tests. Now and then, GANs neglect to demonstrate a multimodal likelihood circulation of information and experience the ill effects of mode breakdown. A circumstance wherein all the produced tests are for all intents and purposes indistinguishable is known as a complete breakdown.

## Future Prospects of GAN

The GAN is in its inceptive phase and there are many directions in which GAN needs to improve. The research on model convergence and existence of equilibrium point is important for improving GANS on future prospects. A type of GAN named WGAN has a better improvement than normal GAN in overcoming the training instability problem and collapse mode problem. There are important research topics like generating data that are capable of interacting with humans from simple random inputs, creating new artificial intelligence applications by integrating GANS with feature learning, reinforcement learning, and imitation learning. The advantages of GAN have opened the doors for new research work in fields like image processing and sequential data computer vision. The main work of GAN was to create feasible synthetic images and had given a very positive response in the field of computer vision, area but its research in other fields is still very limited. The main reason why it has not been able to give a positive response is due to different attributes in image and non-image data. Research is still under progress for use of GAN in the field of natural language processing which is based on discrete value data, whereas GAN works on continuous value data. GANS is used for checking if the samples produced belong to the same distribution with the help of an adversarial discriminator. GAN has been used in the field of statistical parametric speech synthesis for post-filtering where approximating speech parameters involve using DNNs. Here also GAN is used for recreating "spectral texture". Here we pass bad samples as input with noise in the generator, and the output generated by the generator is the cleaned-up text which in comparison with real samples is very similar and is quite different from the distorted sample we gave as input. Generator and discriminator play the adversarial game, and when both of them converge, generator becomes capable of producing samples very much similar to real samples. Research is going on to reducing the time and gaining more stability for generating more polished input. In the field of face generation, GAN has made a quite significant improvement. Many changes in methods from initial phase like improvement of low-quality images by using StackGAN, using better neural network architectures like deep convolutional neural networks for generating faces, solving instability problem of GAN training by using large architectures like ResNet, and generating mega-pixel face image by using multiscale training have shown a path for exploring and researching in this directions. GAN has been also used for generating new and repairing old music, speech, and audio. GAN is also used in many natural language processing applications such as poetry generation, question–answer selection, and review detection and generation. In the field of generating super-resolution images, SR GAN was used for improving photo-realistic natural images, and for improving the visual quality of images, three main key components of SR GAN were studied thoroughly and improvements were made. CycleGAN is also used for generating super-resolution images, and more studies are going on to create high-quality images. The quality of the image has improved year by year due to the rapid progress made by GANs in the recent few years. Much research is going on to reduce the implementation and training of GANs which if reduced can generate sharper and high-quality images.

## Conclusion

In this paper, we saw the use of different GAN which can be used to convert factual descriptions to the human face image. Though there are different types of GANs available, we have seen that not all can be used for text-to-image conversion, and there are still a lot of disadvantages for the one which can be used for the conversion. The state-of-the-art generative adversarial network was discussed in detail. Research directions were provided to improve the details of the image produced by the neural network known as GANs.

# References

1. Xu, T., Zhang, P. Huang, Q., Zhang, H., Gan, Z., Huang, X., He, X.: AttnGAN: fine-grained text to image generation with attentional generative adversarial networks. In: IEEE/CVF Conference on Computer Vision and Pattern Recognition. arXiv:1711.10485 (2017)
2. Odena, A., Olah, C., Shlens, J.: Conditional image synthesis with auxiliary classifier GANs. In: Proceedings of the 34th International Conference on Machine Learning (ICML'17), vol. 70, pp. 2642–2651. JMLR.org (2017)
3. Miyato, T., Kataoka, T., Koyama, M., Yoshida, Y.: Spectral normalization for generative adversarial networks. CoRR arXiv:1802.05957 (2018)
4. Zhang, H., Goodfellow, I., Metaxas, D., Odena, A.: Self-attention generative adversarial networks. In: PMLR, Proceedings of the 36th International Conference on Machine Learning (2019)
5. Li, B., Qi, X., Lukasiewicz, T., Torr, P.: Controllable text-to-image generation. NIPS. arXiv:1909.07083 (2019)
6. Yanagi, R., Togo, R., Ogawa, T., Haseyama, M.: Query are GAN: scene retrieval with attentional text-to-image generative adversarial network. IEEE Access 7, 153183–153193 (2019)
7. Radford, A., Metz, L., Chintala, S.: Unsupervised representation learning with deep convolutional generative adversarial networks. arXiv:1511.06434 (2015)
8. Reed, S.E., Akata, Z., Yan, X., Logeswaran, L., Schiele, B., Lee, H.: Generative adversarial text to image synthesis. ICML. arXiv:1605.05396 (2016)
9. Cha, M., Gwon, Y., Kung, H.T.: Adversarial nets with perceptual losses for text-to-image synthesis. In: IEEE International Workshop on Machine Learning for Signal Processing. Tokyo, Japan, 25–28 Sept 2017
10. Mirza, M., Osindero, S.: Conditional generative adversarial nets. arXiv:1411.1784v1 (2014)
11. Azadi, S., Pathak, D., Ebrahimi, S., Darrell, T.: Compositional GAN: learning image-conditional binary composition. arXiv:1807.07560v3 (2019)
12. Isola, P., Zhu, J.Y., Zhou, T., Efros, A.A.: Image-to-image translation with conditional adversarial networks (2016)
13. Eitz, M., Hays, J., Alexa, M.: How do humans sketch objects? In: SIGGRAPH (2012)
14. Hwang, S., Park, J., Kim, N., Choi, Y., Kweon, I.S.: Multispectral pedestrian detection: benchmark dataset and baseline. In: CVPR (2015)
15. Doersch, C., Singh, S., Gupta, A., Sivic, J., Efros, A.: What makes Paris look like Paris? ACM Trans. Graph. 31(4) (2012)
16. Park, H., Yoo, Y., Kwak, N.: MC-GAN: multi-conditional generative adversarial network for image synthesis (2018)
17. Joseph, K.J., Pal, A., Rajanala, S., Balasubramanian, V.N.: C4Synth: cross-caption cycle-consistent text-to-image synthesis. In: 2019 IEEE Winter Conference on Applications of Computer Vision
18. Gorti, S.K., Ma, J.: Text-to-image-to-text translation using cycle consistent adversarial networks. arXiv:1808.04538v1 (2018)
19. Wan, L., Wan, J., Jin, Y., Tan, Z., Li, S.Z.: Fine-grained multi-attribute adversarial learning for face generation of age, gender, and ethnicity. In: International Conference on Biometrics (ICB), Gold Coast, QLD, pp. 98–103 (2018)
20. Bao, J., Chen, D., Wen, F., Li, H., Hua, G.: Fine-grained image generation through asymmetric training. IEEE. arXiv:1703.10155 (2017)
21. Zhang, H., Xu, T., Li, H., Zhang, S., Wang, X., Huang, X., Metaxas, D.N.: StackGAN++: realistic image synthesis with stacked generative adversarial networks. arXiv:1710.10916v2 (2017)
22. Zhan, F., Zhu, H., Lu, S.: Spatial fusion GAN for image synthesis. CVPR arXiv:1812.05840 (2018)
23. Karatzas, D., Shafait, F., Uchida, S., Iwamura, M., Mestre, S.R., Mas, J., Mota, D.F., Almazan, J.A., de las Heras, L.P., et al. ICDAR 2013 robust reading competition. In: ICDAR, pp. 1484–1493 (2013)

24. Karatzas, D., Gomez-Bigorda, L., Nicolaou, A., Ghosh, S., Bagdanov, A., Iwamura, M., Matas, J., Neumann, L., Chandrasekhar, V.R., Lu, S., Shafait, F., Uchida, S., Valveny, E.: ICDAR 2015 competition on robust reading. In: ICDAR, pp. 1156–1160 (2015)
25. Mishra, A., Alahari, K., Jawahar, C.V.: Scene text recognition using higher-order language priors. In: BMVC (2012)
26. Wang, K., Babenko, B., Belongie, S.: End-to-end scene text recognition. In: ICCV (2011)
27. Phan, T.Q., Shivakumara, P., Tian, S., Tan, C.L.: Recognizing text with perspective distortion in natural scenes. In: ICCV (2013)
28. Risnumawan, A., Shivakumara, P., Chan, C.S., Tan, C.L.: A robust arbitrary text detection system for natural scene images. Expert Syst. Appl. **41**(18), 8027–8048 (2014)
29. Bodla, N., Hua, G., Rama, C.: Semi-supervised fused GAN for conditional image generation. In: 15th European Conference. Proceedings. Part V. Munich, Germany, 8–14 Sept 2018. https://doi.org/10.1007/978-3-030-01228-141

# Chapter 2
# Integration of Machine Learning in Education: Challenges, Issues and Trends

**Salwa Mohammed Razaulla, Mohammad Pasha, and Mohd Umar Farooq**

## Introduction

Education is an essential part of our life that is an ongoing, never ceasing process. Right from the early classroom days at a school to a gradual drift toward eLearning, many techniques and approaches are being developed to produce an enriching experience within the whole educational field. But the content-centered programmes and lecture-based pedagogy are unable to meet the challenges of the new digital age and the requirements of diverse learners. As such, the teaching and learning styles, techniques and approaches are getting influenced by innovations in technology. In order to meet the demands of this new technology-driven world, the curriculum is moving away from content-centered learning and toward a process-centered learning. One example of such a movement is the Next Generation Science Standards (NGSS), developed to improve science education by creating a research-based, up-to-date standard in the subjects of science for all K-12 students. Some of the technology that has been impacting and will continue to impact the teaching and learning processes include robots, machine learning, artificial intelligence, deep learning, learning analytics, natural language processing, AR/VR/3D and cloud. According to EdWeek, AI remains one of the main advanced education technology expenditures after AR/VR. As shown in Fig. 2.1, the expenses spent on AI in education are expected to reach 6.1 billion dollars by 2025.

Machine learning in particular has been making waves in the educational sector with its applications in the field which include modeling intelligent tutoring systems for students, making recommendations about possible future career paths, improving the curriculum, accurately predicting grades of students among others. Lately, educators and learners have converged in response to the demands of the digital age.

S. M. Razaulla · M. Pasha (✉) · M. U. Farooq
Muffakham Jah College of Engineering & Technology, Hyderabad, India
e-mail: mdpasha@mjcollege.ac.in

© The Author(s), under exclusive license to Springer Nature Singapore Pte Ltd. 2022    23
Ch. Satyanarayana et al. (eds.), *Machine Learning and Internet of Things for Societal Issues*, Advanced Technologies and Societal Change,
https://doi.org/10.1007/978-981-16-5090-1_2

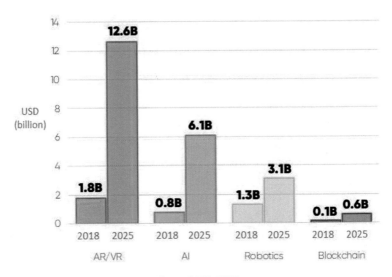

**Fig. 2.1** Advanced technologies expenditures 2018–2025

But while implementation of machine learning with education has the potential to enhance the learning, educators must understand the implications machine learning systems will have on education. This increasing influence raises some concerning questions: who will be teaching whom, what roles do machines and humans play in the learning process? What happens to the traditional roles in the classroom? Are teachers and lecturers going to be rendered obsolete? To address these concerns, in October of 2019, about 150–200 people came together to attend the EduSummIT in Quebec City, Canada. Edu SummIT is a group of policy-makers, practitioners and researchers who are dedicated to effectively integrating information technology (IT) into the education sector by promoting active dissemination and use of research. The purpose of this meeting was to consider the misalignments due to constantly changing knowledge representations, human–computer interactions and plenty of other emerging IT influences. These developments call for new alignments to be made between conventional and modern curricula, between learners and teachers and between learning and evaluation. A summary Action Agendas of the 12 thematic working groups of this meeting was published in [1].

## Machine Learning Overview

Machine learning, a term coined in 1959 by Arthur Samuel, an American pioneer in the field of computer gaming and artificial intelligence, is defined as a field of study that gives computers the ability to learn without being explicitly programmed [2]. In 1997, Tom Mitchell gave a definition that is more widely used today—"A computer program is said to learn from experience E with respect to some task

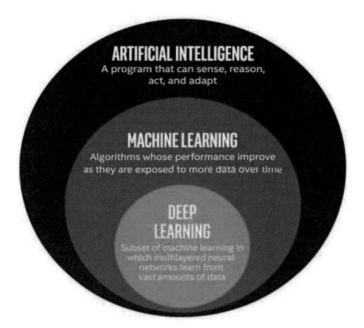

**Fig. 2.2** Artificial intelligence and its subsets

T and some performance measure P, if its performance on T, as measured by P, improves with experience E" [3]. The main goal of machine learning (ML) is to construct computer programs which are capable of learning from the data that they are provided. Machine learning also coincides with numerous other research fields such as mathematics, neuroscience, cognitive science and information theory among many others. Machine learning is considered a subfield of artificial intelligence, as shown in Fig. 2.2. Before the inception of machine learning, AI programs were capable of only automating low-level tasks such as straightforward rule-based classifications or intelligent automation. An example is a rule-based system like chatbots which answer questions and assist customers with human-defined rules but to a limited extent. This implied that AI algorithms were confined to the field they were programmed for. However, with machine learning, AI systems were able to develop beyond just performing the tasks they were programmed for and begin evolving with every cycle. The main distinction that sets machine learning apart from AI is that machine learning has the capability to evolve. The way machine learning algorithms do this is by processing large amounts of data and improving over previous iterations by learning from the data they are provided. So to speak, big data is one of the most significant features of machine learning algorithms, and its efficiency is contingent on the standard of the dataset provided to it.

## Opportunities of Machine Learning in Education

Machine learning can adapt to learning processes to provide personalized learning, recommend courses and other useful related information based on a student's goal, predict student's grades, provide feedback through student interaction and measure student's engagement, etc. According to the US National Center for Education Statistics, nearly 30% of students studying in universities are likely to pursue a course on online platforms every semester. These online platforms integrate machine learning in education to guide learners in their studies. As shown in Fig. 2.3, a growth of nearly 13% compound annual growth rate (CAGR) is expected to be seen through the period of forecast, increasing the global corporate eLearning market to $30 billion.

Assessment or correction of papers has been the primary method of evaluating student's performance, and so it is a vital part of the education process. Manual correction is a laborious and time-consuming task and therefore expensive. But machine learning can automate this process. Li proposes a supervised model to provide a thorough analysis of exam papers to not only correct papers but also help teachers understand the causes that affect a student's learning performance [4]. The model is built using a combination of regression algorithm and several machine learning data analysis methods, such as K-means, deep neural network (DNN) among others in order to bolster the prediction accuracy of learning. A large collection of educational data such as individual information, learning abilities, study habits, education level and professional background is used by regression method. This method can then find the chief determining factors that have lasting impacts on the learning performance. In order to predict the most important criteria which can affect the prediction result, DNN employs this prior knowledge in order to build a training

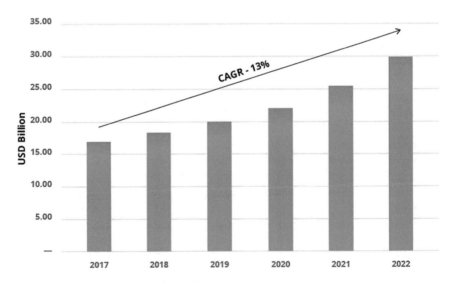

**Fig. 2.3** Global corporate eLearning market rate

model. Students are clustered according to their scores using the K-means method. Many different machine learning models like supervised learning, natural language processing (NPL), etc., are being used in existing learning and eLearning approaches either independently or along with crowdsourcing [5].

One factor that is directly linked to student's productivity and learning gain is engagement or learning engagement. It is a significant task for educators to detect engagement status of their online learners accurately and efficiently in order to deliver personalized pedagogical assistance for them [6]. The benefits of fostering learner's engagement can be felt beyond the online learning as it can also impact other learning platforms such as traditional classrooms, educational games and intelligent tutoring systems [7]. Log files are created to preserve the data collected on student's actions and behavior and various other traits. Such data is extracted from images captured by the image sensors, neurological sensors or by tracing students' activities in their studying environments. These logs are analyzed using different machine learning as well as data mining approaches, to provide useful information for engagement detection [6]. Booth et al. measured learner's engagement in both a scalable and accessible manner [8].

In their study, learners studying online lecturer videos were observed using a screen-mounted camera, and later, these videos were analyzed. Facial landmarks, eye gaze, emotion probabilities, average optical flow magnitude and direction, head pose and size were used along with action units (AUs), also known as facial muscle movements. Finally, by employing the K-nearest neighbor (KNN) classifier, a conclusion was drawn on the engagement detection. Another application that machine learning can find in education is improving the curriculum design. Some of the factors which influence the design of a course module include personal experiences of the faculties, academic paths of students among many others, and as such, disagreements among faculty members are likely to emerge regarding which curriculum best aides the students. Ball proposes using machine learning to improve curriculum design by analyzing student data [9]. This data may include age, gender GPA, and transfer credits, etc., that are taken from a student's transcript. The paper details a two-step process. For the first step, two machine learning algorithms, namely decision trees and logistical regression, are used, in order to quantify the impact of each individual feature on graduation. Secondly, in order to achieve higher graduation rates, the results of the analysis from the first step are used as the premise for applying requisite changes to the course programme.

One optimistic approach is using machine learning in combination with other technologies like learning analytics (LA), education data mining (EDM), etc. Learning analytics can be defined as a set of steps that are followed in order to understand and optimize the entire learning process, along with the environment in which the learning occurs. An intelligent LA generally deals with a large amount of educational data, and using machine learning algorithms has proven to be beneficial because of its ability to analyze big data and produce reliable, repeatable results by learning from the previous combinations. One of the major tasks of LA is predicting the test outcomes. Kangwook et al. proposes using the collaborative filtering algorithm over logistic regression because it is a completely data-driven method, which means that it does not require any extra information apart from the test response data [10]. In this paper, two models are proposed, a new model for test responses and collaborative filtering algorithm, which has increased human-interpretability based on this new model? The prediction capability of these approaches is evaluated by using an extensive educational dataset, which is accumulated via mobile applications. The results of their experimental results demonstrate that the collaborative filtering approach is far better at predicting test outcomes than the logistic regression approach.

## Challenges and Issues

As we move into a digitally dominated world, emphasis should be given to the possible challenges as well as opportunities that arise as educational technologies additionally enhance with the inclusion of AI and its subsets. Some of the challenges and issues to consider are listed below:

### *Explainability*

The decisions and predictions a machine learning system makes function as a black box due to the fact that these models are extremely complex and inexplicable by nature and as such hinder verification of the reasoning processes. But there are many situations where the operational characteristics together with the data used should be transparent so that decisions and predictions made by algorithms can be justified. This transparency is required to reduce bias and make certain that decisions and conclusions made by machine learning algorithms are fair, interpretable by all. It is hard to detect such damage caused by algorithmic activity and is even harder to find the root cause. Additionally, it is seldom straightforward to find the person or persons responsible for any such harm caused due to sheer number of people involved in designing and developing such complex systems.

Webb et al. describe two major issues of using machine learning in education— explainability and accountability [11]. Explainability is described as the ability to understand and explain "in human terms" what is happening with the model; how exactly it works under the hood. Although not every situation requires it, some level of explainability for machine learning systems need to be employed in education, except in cases where there are no consequential outcomes of delivering incorrect output results. It need not be explained that machine learning systems developed for the use of students will have a significant impact for them in the educational area, but there are some systems that can have extraordinary consequences in situations of high stake evaluations. Creating models to explain these black box systems in hopes that it will mitigate some of these issues does not only account for a bad practice but can also end up causing irreparable damage to the society. Recently, there has been a lot of research work on "Explainable machine learning/Explainable

AI (XAI)" where a second model is created afterward to explain the first black box model. The key issues with explainable machine learning are that these models (i) have algorithms/functions that are too complicated (ii) the function is proprietary, and thus, the working and details are kept a secret. Rudin proposes developing models that are intrinsically interpretable instead, as they supply their own rationalizations, which are faithful to what the model actually evaluates [12].

## *Accountability*

Because machine learning systems are capable of learning independently without human intervention, this raises the issue of fairness, transparency, privacy, data security, more commonly bundled together as "algorithmic accountability". Accountability is defined as the ability to provide reasons with the aim of explaining and justifying methods, actions and decisions. Since the advent of digital technology, a lot of decisions we make are based on algorithms of these systems, and thus, accountability regarding the logic, processing and deployment of these algorithms are necessary. Evaluation of machine learning components, the weights and biases used as learning parameters as well as any other element that might have influenced the algorithmic outcome are all imperative to having a clear account of an automated decision. The designer of the algorithm too needs to be held accountable for the outcome of the decisions and judgment the system makes. Unfortunately, such detailed evaluation of the algorithmic decision-making can be challenging due to the following reasons: (i) they are complex in nature; often viewed as a black box, i.e., they do not allow analysis of the decision-making technique and its end results (ii) operations based on random group level are very common in algorithms which can hinder the attempted interpretation of its risks and processes. Algorithms tend to be complex in at least two aspects: contextually and technically.

The technical complexities are associated with the nature of algorithms, such as its mathematical constructions. They are contextually and relationally complex in relation to the performance of tasks with data and input with users under various settings. There is an obvious difference between the decisions made by a human, who think rationally and emotionally taking into consideration the moral and ethical values, drawn from experiences, judgments and individual flexibility whereas a machine learning algorithm, which bases its outputs on the given available input data to provide "sheer facts and truths". Unfortunately, the common belief that "computers don't make mistakes" overlooks the fact that data and algorithms can be easily designed to be biased or that algorithm functions value-driven themselves.

## *Cultural Bias*

AI-enabled educational technology has a global audience and wide range of users. It should thus take into consideration the challenges of cultural inclusiveness. Its extent should not only be concerned with students with special needs but at the same time with different cultural groups who should be provided with quality and completely accessible education. In this era of digital age, it is important to tackle inclusiveness concerning the probable cultural biases in technology-powered education [13]. It is crucial to consider the cultural backgrounds of learners when designing and modeling machine learning systems because a person's culture shapes their perception of the world, character, values and aspirations and also how they respond to technology-driven learning [14]. This cultural bias in predictive and decision-making systems might, for example, be due to the biased training datasets which hold information with the intentional, systematic or political discrimination, or certain data which could have been gathered from a selected demographic that does not represent the entire population. The datasets could also hold implicit gender biased, racial or ideological content. Designing a learning environment with such datasets will possibly result in exclusion of certain groups of learners. Some data collection techniques used for training the machine learning algorithms to create an adaptive learning environment include facial recognition and speech recognition. However, research shows that majority of the facial recognition systems is sensitive and discriminative toward people of color [15].

A language barrier could also be a hindrance for the speech recognition software. In order to include the myriad of student groups, the implementation of cultural and racial inclusion in educational technologies needs to be evident to allow full participation of users without stereotyping gender, race and ethnicity and to prevent undesirable associations. Kazimzade et al. describe some steps that can be taken to avoid bias in educational technologies [16]. The first step is to investigate the datasets that are used to train the training model. Care should be taken to ensure that the training data contains equitable samples in the database and whether this training data was tested on users who were not part of the training examples. There should also be an emphasis on the team that is developing these algorithms and ensure it is a diverse group that is capable of recognizing biases. Another factor that could contribute to a lack of diversity is if unknown individual patterns are incorrectly regarded as a selection criterion for existing popular patterns. AI systems need to be designed in a manner that allows dynamic evolution as its users, teachers or students change over time.

## *Ethical Concerns*

The meteoric rise in the use of digital age computing in education has not been a slow development. This dominant inclination toward the importance and significance of technology on education has had a strong influence on educators and their teaching methods, as they are required to merge the developments of learner's basic skills as well as analytical competence and creativity in a globally technological and economically challenging society [17]. Nevertheless, this skills-based pedagogy together with continually rising technologies have obscured the learner's personal creativity, intellectual and collaborative learning, and the sense of consideration and sensitivity, which has devalued the original goal of education [18]. The argument remains that technologization of education has had a tremendous effect on the learning process because of its focus on education as Erziehung, or education as the learning of a skill or trade, to the detriment of education as Bildung, or education as character formation [19].

The author goes on to point out that this focus of education as Erziehung has dire implications on the ethical, social and political aspects of education because it interferes directly and negatively with the individual's capacity to be someone who is concerned for others in the community, who engages with the various problematic issues of society, and who is aware of the impact of their actions upon themselves and society as a whole. And what makes matters worse is that this interference and possible depersonalization of the classroom is not questioned or scrutinized in greater detail. Society has been too eager to adopt the success of technology because of the ease it provides in measuring and analyzing student performance and progress [18]. Erziehung along with a pedagogical shift from teaching to learning has made an environment that promotes the process of "learnification", coined by Gert Biesta, which encourages the notion that the creation of rich learning environments should be the primary concern of teaching. This is supported by the use of technological aids in classrooms for the purpose of instructional scaffolding.

## Existing Examples

Duolingo is an instance of different ways in which deep learning is being utilized in the learning context, through assistive tutoring. It is an AI-based language learning application. In order to make intelligent predictions, Duolingo uses neural networks to mimic the human brain. Using machine learning algorithms, the application analyzes user log data to predict the difficulty level of any given exercise for every individual user and the likelihood that they will get an answer correct. These predictions enable Duolingo to be the ultimate personalized learning system. To develop active learning, the system runs a natural language processing (NPL) for every correct answer and a NPL pipeline for every wrong answer, and the differences between these two are compared to come up with explanations of what is wrong. There also is an option for users to appeal when an answer they submitted was deemed wrong by hitting the report button. But not all these reports are legitimate since some of these can be accidental taps or the user's answer is indeed incorrect, but they might be unaware of it. The application also has a built-in machine learning system that uses logistic regression algorithm that would surface only the useful reports that were filed by users.

The University of Michigan developed a tool called M-Write to help improve the writing of students in large enrolment courses in multiple departments of the university. The M-Write program uses a machine learning algorithm to analyze the tradeoffs of writing tasks and identifies students who are lagging behind. Vocabulary matching and topic matching are some techniques that the algorithm uses to analyze student's texts. Information such as reviews and scores from professors and fellows is collected from students who had previously been using the M-Write tool. The developers use this data to build the automated system that can recognize students who are having difficulty with the study materials.

Another university that has integrated machine learning is the University of California. The Computer Science department's professor Pavel Pevzner and his colleagues developed an advanced online course for undergraduates designed specifically as an adaptive intelligent tutoring system (ITS) for the edX platform. Through this platform, those who are enrolled in the MOOCs receive an adaptive and personalized learning path. The students are continuously evaluated during the course, through various quizzes and tests, especially the "just in time" exercises that allow for individual basement. If systems such as ITS are applied in the education sector, it would significantly alter the traditional methods of teaching and learning to instead take course toward a mastery-based learning platform.

Another example is Quizlet, an online studying platform, that lets users create their own quizzes, flashcards or use pre-existing ones in the database. Using a combination of machine learning and statistics, the team leverages user data to discover how students can study more effectively. The platform's Learning Assistant algorithm takes factors such as direction of study, correctness of an answer and time between previous answer to prioritize words that students are most likely to forget, ultimately helping them learn more effectively. Learning Assistant combines machine learning data with past study performance.

## Conclusion

The progress machine learning has brought to businesses, health care, agriculture, finance has also the potential to provide a new direction in the field of education. Although these systems might lack the emotional and rational thinking that is a characteristic of humans, their predictive ability is unparalleled. As the world inches closer to a fully AI-powered automation, it is clear that machine learning systems will be faster, more precise at making inferences and predictions than humans. However, in order to employ these machine learning-based systems meaningfully to support the learning environment raises the issue of their explainability, interpretability and accountability [20]. In this chapter, we discussed some of these issues and how to overcome them. We also presented various opportunities machine learning can find in the context of teaching to help improve the standards of the learning process.

Finally, we list out some scenarios of how companies and educational institutions are employing machine learning to design innovative and more intuitive methods for learning.

# References

1. Wang, Y., Gallagher, M.: Artificial Intelligence and Inclusive Education. Perspectives on Rethinking and Reforming Education. Springer, Singapore. https://doi.org/10.1007/978-981-13-8161-4_10
2. Samuel. A.L.: Some studies in machine learning using the game of checkers. II—recent progress. In: Levy, D.N.L. (eds.) Computer Games I. Springer, New York, NY (1988). Res. Develop. **44**(1.2), 210–229 (1959). https://doi.org/10.1007/978-1-4613-8716-9_15
3. Mitchell, T.: Machine Learning. Mc Graw-Hill International Editions (1997)
4. Li, Y.F., Liang, D.M.: Safe semi-supervised learning: a brief introduction. Front. Comput. Sci. **13**, 669–676 (2019). https://doi.org/10.1007/s11704-019-8452-2
5. Alenezi, H.S., Faisal, M.H.: Utilizing crowdsourcing and machine learning in education: literature review. Educ. Inform. Technol. **25**, 2971–2986 (2020). https://doi.org/10.1007/s10639-020-10102-w
6. Dewan, M.A.A., Murshed, M., Lin, F.: Engagement detection in online learning: a review. Smart Learn. Environ. **6**, 1 (2019). https://doi.org/10.1186/s40561-018-0080-z
7. Karumbaiah, S., Woolf, B., Lizarralde, R., Arroyo, I., Allessio, D., Wixon, N.: Addressing student behavior and affect with empathy and growth mindset. International Conference on Educational Data Mining, Wuhan (2017)
8. Booth, B.M., Ali, A.M., Narayanan, S.S., Bennett, I., Farag, A.A.: Toward active and unobtrusive engagement assessment of distance learners. International Conference on Affective Computing and Intelligent Interaction, San Antonio (2017)
9. Ball, R., Duhadway, L., Feuz, K., Jensen, J., Rague, B., Weidman, D.: Applying machine learning to improve curriculum design. In: Proceedings of the 50th ACM Technical Symposium on Computer Science Education (SIGCSE '19). Association for Computing Machinery, New York, NY, USA, 787–793 (2019). doi:https://doi.org/10.1145/3287324.3287430
10. Lee, K.: Machine learning approaches for learning analytics: collaborative filtering or regression with experts? Korea, 1–11 (2018)
11. Webb, M.E., Fluck, A., Magenheim, J., et al.: Machine learning for human learners: opportunities, issues, tensions and threats. Educ. Tech. Res. Dev. (2020). https://doi.org/10.1007/s11423-020-09858-2
12. Rudin, C.: Stop explaining black box machine learning models for high stakes decisions and use interpretable models instead. Nat. Mach. Intell. **1**, 206–215 (2019). https://doi.org/10.1038/s42256-019-0048-x
13. Blanchard, E.G.: Is it adequate to model the socio-cultural dimension of e-learners by informing a fixed set of personal criteria? In: 12th IEEE International Conference on Advanced Learning Technologies (ICALT), USA, pp. 388–392 (2012)
14. Collis, B.: Designing for differences: cultural issues in the design of WWW-based course-support sites. Br. J. Educ. Technol. **30**, 201–215 (2002). https://doi.org/10.1111/1467-8535.00110
15. Buolamwini, J.: Gender shades: intersectional phenotypic and demographic evaluation of face datasets and gender classifiers (2017)
16. Kazimzade, G.: Artificial Intelligence in Education Meets Inclusive Educational Technology—The Technical State-of-the-Art and Possible Directions. https://doi.org/10.1007/978-981-13-81614_4
17. Turner-Smith, A., Devlin, A.: E-learning for assistive technology professionals—a review of the telemate project. Med. Eng. Phys. **27**, 561–570 (2005)

18. Laura, R.S., Chapman, A.: The technologisation of education: philosophical reflections on being too plugged in. Int. J. Children's Spirituality **14**(3), 289–298 (2009). https://doi.org/10.1080/13644360903086554

19. Guilherme, A.: Considering AI in education: Erziehung but never Bildung. In: Knox, J., Hodges, J., Mohan, S. (eds.). Machine learning in gifted education: a demonstration using neural networks. Gifted Child Q. **63**, 243–252 (2019). https://doi.org/10.1177/0016986219867483.

20. Weller, A.: Challenges for transparency. In: Proceedings of the ICML Workshop on Human Interpretability in Machine Learning, pp. 55–62 (2017)

# Chapter 3
# IoT-Based Continuous Glucose Monitoring System for Diabetic Patients Using Sensor Technology

**Anchana P. Belmon and Jeraldin Auxillia**

## Introduction

A dynamic network with all the physical and virtual objects interconnected can be viewed as the mainstream known as Internet of things (IoT). The main streams of IoT, namely cloud computing, artificial intelligence and wireless sensor networks can be viewed as an important role comprising health care, transportation and logistics. An innovation in e-health and application in wellness paved a vital role in nutrition, fitness and wellness application systems. Fully autonomous health-based wireless monitoring systems benefit mostly to ageing population with many useful applications. According to WHO, the major cause of mortality is due to diabetes with almost exceeding 1.5 million people per year. Diabetes mellitus is a major chronic disorder mainly caused due to the pancreas not producing enough insulin or the body not producing enough insulin. Pancreas produce hormones called as insulin that control the level of glucose to fix the various needs of tissues and organs. The insulin takes advantage of using nutrients from the body's carbohydrate food. The deficiency in the processes causes significant impact in the glucose accumulation leading to human health deterioration.

Diabetes mellitus seriously affects the personal and societal well-being of persons. Although there is no permanent cure for diabetes, the main solution is to continuously measure blood glucose levels and close it with an appropriate delivery of insulin. As per the UK prospective diabetes group, the continuous glucose monitoring can reduce the diabetic complications at about 40–75%. Hence, for corrective actions regarding diet, physical exercise and medication, continuous glucose monitoring equipped with alarm systems is highly preferred to reduce the complications.

A. P. Belmon (✉)
Maria College of Engineering and Technology, Attoor, India

J. Auxillia
St. Xaviers Catholic College of Engineering, Chunkankadai, India

© The Author(s), under exclusive license to Springer Nature Singapore Pte Ltd. 2022
Ch. Satyanarayana et al. (eds.), *Machine Learning and Internet of Things for Societal Issues*, Advanced Technologies and Societal Change,
https://doi.org/10.1007/978-981-16-5090-1_3

Wireless sensor applications are highly preferred by the patients and healthcare providers with suitable implications on the sensor imparted causing pain in imparting battery. For fully automated energy harvesting applications, it is cautious to design the electronic circuitry with less power and more efficient energy source extraction schemes.

## Glucose Measurement Methods

The blood glucose parameter control is the main concern with the people of diabetes mellitus. Blood glucose control parameters beyond boundaries are monitored under three main categories, namely (1) Invasive (2) Minimally invasive (3) Non-invasive. The commonly used techniques are the invasive methods as they offer greatest precision due to the close contact with the patient's blood. Piercing a finger in a clean regimen is the traditional procedure which seems to be painful. The minimally invasive technique uses micropores (small holes) on the skin using the radiations from LASER. Through micropores, transdermal body fluids are extracted using continuous vacuum pressure. In non-invasive measurement, glucometers are usually preferred in external means by fixing a sensor in the specific region of the body. The amount found under the skin is taken as the blood glucose amount with a good reliability. The other biofluids such as saliva, sweat, urine and tears are considered as the non-invasive glucose tests without continuous glucose tracking. Continuous glucose variations throughout the body improve the metabolic control of the diabetic patients improving the glucose variations throughout the whole day. The critical values can be avoided by the prediction of values. Recent approaches include monitoring patient's glucose parameters. This non-invasive method includes transmission technology and also the processing of data with timely readings.

Glucometer-based blood glucose measurement is the invasive method. Recent advancement in the invasive blood glucose measurement involves (1) **Alternative measure** with painless blood sampling and the test strip usage. (2) **Multi-test systems** use multiple test strips. (3) **Uncoded systems** use older test strips with error detection and measurable values. The risk of detecting errors can be measured by 'auto coding and single coding'. If each test strip is corrected, then it is called auto coding, whereas similar code correction for all test strips is done in single coding. (4) **Recordable parameters** use software to measure and store results.

In non-invasive methods, sensors are targeted to skin, ear lobe areas. The various non-invasive approaches are (1) **Near Infrared (NIR) Spectroscopy:** NIR spectroscopy utilizes infrared range of the electromagnetic spectrum (800–2,500 mm) which penetrates deeply in the samples. (2) **Ultrasound Technology:** Ultrasound-based technology with a frequency greater than 20 kHz and it can elaborate the structure of object under investigation. (3) **Dielectric Spectroscopy:** The dielectric spectroscopy helps to measure the dielectric properties of all the objects based on the frequency. (4) **Metabolic Heat Confirmation:** The metabolic heat confirmation is the method based on the degree of oxygenation in the blood. It measures the amount of heat dissipated.

Non-invasive types of methods are time-consuming processes as the doctor has to meet the patient; however, the food diet and the control of exercise are impossible.

## Existing Methods

The non-invasive techniques find the blood as a substitute for glucose to make the patient with system portable. The first effort [1] was based on the detection in resonance-related fluorescence energy with a protein efficiency for analytic concentrations. For glucose monitoring, advanced technologies like GSM [2] are probably used. Many different varieties of sensors made non-invasive are nowadays available with a good sensing operation. Some of the examples include zinc oxide-based nanowires, GaAs-related infrared emitter diodes. The research community focuses on the glucose levels to form a comprehensive system. The cardiac type diseases were also considered besides these glucose levels. The wireless communication [3] is done on the Zigbee [4], and the display readings are noted. For monitoring applications, real-time low power-based wireless sensor networks (WSNs) shows an attractive property. To reduce the size and improve portability, WSN is used in conjunction with invasive sensor [5] applications. The sensor is placed inside the body's subcutaneous tissue layer to determine the oxygen and the amount of glucose. The conditions of elderly people make portability as the major factor. Simple battery free devices are made with the use of sensors and ASIC designs. Also, for the connection and visual impact-based tasks, smartphones and Android operating system characteristics were used. To communicate with mobile devices, a collection of invasive and non-invasive sensors are used. The patient can even monitor the readings, and it can be further stored in a cloud for the view of patient's doctor and relatives.

A glucose monitoring system is normally a Low Energy Bluetooth technology (LEB). Data collected based on the glucose is transmitted from LEB to smartphone for visualization. This system reduces power consumption [6–9] of the implantable unit. A long-term sensor implantable [10, 11] system monitors glucose data for every 2 min to the receivers externally. This continuous usage system can be implanted 180 days in the human body for managing diabetes mellitus. A non-invasive glucose monitoring system [12–14] with near infrared (NIR) depends on the NIR sensor [11, 15, 16]-based received signal intensity. A better wireless network of blood glucose monitoring systems for diabetes detection [11] monitors glucose on wide area networks which predicts the future glucose levels also.

## Internet of Things (IoT)

The connection of all physical devices will help to send and receive information about patients through the internet. The system, devices and related services are enhancive connected through IoT. The automation in those applications requires internetworking these devices. As per the current statistics, studies in 2020 by Gartner over 20.8 billion devices are to be connected to IoT. Remote health condition monitoring and notification intimation at emergency are based on IoT.

## General IoT-Based Continuous Glucose Measurement Architecture

A general IoT-based continuous glucose measurement architecture is depicted in Fig. 3.1. The mandatory sensors, namely glucose sensor, temperature sensor, pressure sensor, pulse sensor, are present in the sensor layer. These sensors are connected to an Arduino nano-microcontroller. The Arduino nano-microcontroller is connected with the Raspberry Pi and cloud through WiFi in the receiver side. The Arduino nano-microcontroller converts the analog readings from the sensor to the digital readings.

In the transmitter section as in Fig. 3.2a, the patient readings are transferred to the Raspberry Pi. The glucose sensor (one touch glucometer) is simple and highly efficient. The colour alerts and audio alerts specify the increase in glucose levels. The receiver section (Fig. 3.2b) collects the patient information using WiFi through the Raspberry Pi and stores the collected data into a cloud. An SD card as a mass storage like a hard disk is incorporated into the Raspberry Pi board. The bootable Linux operating system in the card supports a Raspberry Pi of Linux, ARM and Mac Operating systems. By selecting one OS, we can write an SD card using disk manager application. The Raspberry Pi-enabled WiFi uploads all patient data readings into the cloud.

**Fig. 3.1** General IoT-based continuous glucose monitoring system for diabetes patients

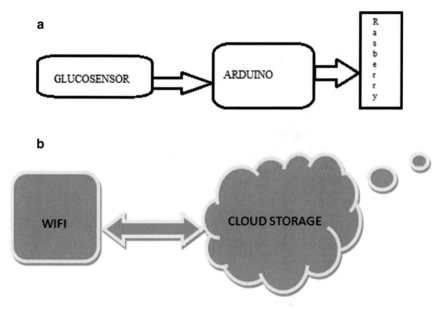

**Fig. 3.2** **a** Transmitter section. **b** Receiver section

## Internet of Things (IoT)-Enabled Continuous Glucose Monitoring System (CGMS)

IoT applications are used to provide information to the diabetic patients to take preventive measures such as diet and exercise. The working principle of the glucose measure is done by the glucometer kit. The server connected to the kit through USB mode stores the measured glucose value. The database as in Fig. 3.3 in the server consists of three basic levels, namely: low, normal and high level. Low diabetic level in the database is between 60 and 80 mg/dl. Normal diabetic level [17] should be maintained between 81 and 140 mg/dl. While the high diabetic level is the blood glucose level above 140 mg/dl. The serial mode of communication is enabled through the USB cable. The server also provides information regarding the diet control and exercise control to maintain glucose levels. After all, the information retrieved from the server database is sent by SMS to the doctor and the user. The usage of the glucometer kit and the IoT exploitation helps to monitor the diabetic patient lifestyles using an automated pervasive system without doctor consultation.

## Results and Discussion

The readings of glucose for different timings like fasting, before and after meal, bed time and exercise are taken, and the variations are noted as a measurement of mg/dl.

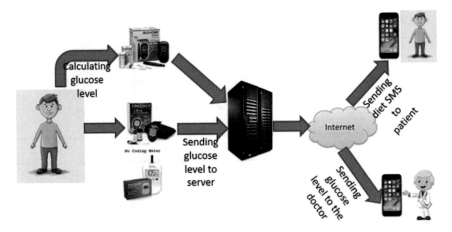

**Fig. 3.3** Example of IoT-based continuous glucose monitoring system

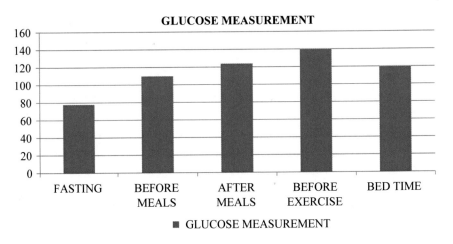

**Fig. 3.4** Graph showing glucose measurement for different timings

The measured results are analysed and plotted in a graph. The resulted values of glucose levels for a single patient measurement are found to be accurate at all the intervals of time as in Fig. 3.4.

## Conclusion

Continuous health monitoring of diabetic patients is costlier nowadays. Using IoT technology, doctors can monitor at any time outside the hospital after consulting hours. To get a better clinical outcome, resources are exploited by health diagnosis smart devices. Also, the IoT-based applications method reduces bed days and stay length in hospitals.

# References

1. Blood Sugar Level Ranges. Diabetes.co.uk. https://www.diabetes.co.uk/diabetescare/blood-sugar-level-ranges.html. Accessed 22 Dec 2016
2. Menon, K.A.U., et al.: A survey on non-invasive blood glucose monitoring using NIR. In: ICCSP 2013, pp. 1069–1072. IEEE (2013)
3. Murakami, A., et al.: A continuous glucose monitoring system in critical cardiac patients in the intensive care unit. In: 2006 Computers in Cardiology, pp. 233–236. IEEE (2006)
4. Aragues, A., et al.: Trends and challenges of the emerging technologies toward interoperability and standardization in e-health communications. IEEE Commun. Mag. (2011)
5. King, P., et al.: The UK prospective diabetes study (UKPDS): clinical and therapeutic implications for type 2 diabetes. British J. Clin. Pharmacol. (1999)
6. WHO. Global report on diabetes. http://apps.who.int/iris/bitstream/10665/204871/1/978924 1565257eng.pdf. Accessed 22 Dec 2016
7. Ali, M., et al.: A bluetooth low energy implantable glucose monitoring system. In: EuMC 2011, pp. 1265–1268. IEEE (2011)
8. Al Rasyid, M.U.H., et al.: Implementation of blood glucose levels monitoring system based on wireless body area network. In: 2016 IEEE International Conference on Consumer Electronics-Taiwan (ICCE-TW), pp. 1–2. IEEE (2016)
9. Wang, N., Kang, G.: A monitoring system for type 2 diabetes mellitus. In: Healthcom 2012, pp. 62–67. IEEE (2012)
10. Al-Sarawi, S.F.: Low power schmitt trigger circuit. Electron. Lett. **38**(18), 1009–1010 (2002); Ali, M.: Low Power Wireless Subcutaneous Transmitter. PhD thesis (2010)
11. International Commission on Non Ionizing Radiation Protection. ICNIRP guidelines for limiting exposure to time varying electric, magnetic and electromagnetic fields (up to 300 GHz). Health Phys. (1998)
12. Gia, T.N., et al.: IoT-based fall detection system with energy efficient sensor nodes. In: NORCAS 2016, pp. 1–6. IEEE (2016)
13. Sudevalayam, S., Kulkarni, P.: Energy harvesting sensor nodes: survey and implications. IEEE Commun. Surv. Tutorials (2011)
14. Jelicic, V., et al.: Analytic comparison of wake-up receivers for WSNS and benefits over the wake-on radio scheme. In: PM2HW2N '12, pp. 99–106. ACM (2012); Gu, L., et al.: Radio-triggered wake-up capability for sensor networks. In: RTAS 2004, pp. 27–36 (2004); Kuan-Yu, L., Tsang, T.K.K., Sawan, M., El-Gamal, M.N.: Radio-triggered solar and RF power scavenging and management for ultra low power wireless medical applications. In: 2006 IEEE International Symposium on Circuits and Systems, pp. 4–5731 (2006)
15. Lucisano, J., et al.: Glucose monitoring in individuals with diabetes using a long-term implanted sensor/telemetry system and model. IEEE Trans. Biomed. Eng. (2016)
16. Taghadosi, M., Albasha, L., Qaddoumi, N., Ali, M.: Miniaturised printed elliptical nested fractal multiband antenna for energy harvesting applications. IET Microwaves Antennas Propag. (2015)
17. Blood glucose monitoring. Diabetes Australia. https://www.diabetesaustralia.com.au/blood-glucose-monitoring. Accessed 22 Dec 2016

# Chapter 4
# Role of Machine Learning and Cloud-Driven Platform in IoT-Based Smart Farming

**Astik Kumar Pradhan, Satyajit Swain, and Jitendra Kumar Rout**

## Introduction

Regardless of environmental challenges such as extreme weather and climate variability, the agricultural industry must grow to meet demand, facing a growing population. Agriculture will need to introduce modern technologies in order to achieve a competitive edge in order to meet the needs of an ever-increasing population. Through IoT, modern agricultural technologies in smart farming and precise farming can help the industry to increase productivity, lower costs, reduce emissions, and improve yield quality. Smart farming is an evolving system that utilizes advanced information communication technologies to manage farms to enhance the quantity and quality of goods while minimizing the necessary human labor. A system for tracking the crop field using sensors (light, temperature, humidity, moisture levels, etc.) and automating the irrigation system is installed in ML and IoT-based smart farming [1]. From anywhere, farm workers can track the field conditions. In comparison with the traditional method, IoT-based smart farming is highly effective. In addition to targeting typical, massive farming operations, IoT-based farm management applications can be new triggers to encourage other expanding or common agricultural trends, such as organic agriculture, household farming (complicated or limited areas, specialized dairy and/or crops, conservation of particular or high-quality varieties, and so on), and to improve highly transparent farming. Smart farming can bring great advantages in terms of environmental problems, along with more effective use of groundwater or management of inputs and treatments. The IoT seeks to simplify processes by eliminating human-to-human interaction [2]. The IoT collects data in the cloud with sensors, the processor processes the data, and the actuators finish the operation. The IoT's target in agriculture is to simplify all aspects of agriculture and methods of farming to make the job very successful. Smart systems collect the data in the cloud-based platform for assessments and deliver reliable results, enabling them to take decisive action. Existing uses of smart farming systems include the collection of data on environmental parameters such as soil water level, air temper-

A. K. Pradhan · S. Swain · J. Kumar Rout (✉)
Kalinga Institute of Industrial Technology, Campus 15, Bhubaneswar, Odisha 751024, India

© The Author(s), under exclusive license to Springer Nature Singapore Pte Ltd. 2022
Ch. Satyanarayana et al. (eds.), *Machine Learning and Internet of Things for Societal Issues*, Advanced Technologies and Societal Change,
https://doi.org/10.1007/978-981-16-5090-1_4

ature, humidity, and pH values [3]. Agricultural ML offers much more reliability, enabling farmers to manage animals and plants almost independently, drastically strengthening the effectiveness of their choices. A broad variety of ML technologies exist for smart agricultural activities which includes yield estimation, water control, and live stock management, to name a few. Smart farming's driving forces are IoT and ML, linking farm-integrated smart machines and sensors to enable agricultural operations data-driven and data-enabled. The data you can extract from stuff ("T") and transfer over the Internet ("I") is at the heart of the Internet of Things [4]. IoT devices mounted on a farm can gather and compile data in a repetitive cycle or a loop to automate the farming process, enabling farmers to respond quickly to new problems and changes in environmental conditions.The following form of a loop as shown in Fig. 4.1 is accompanied by smart farming:

- **Observation**: A system for observing the crop field with the aid of sensors (humidity, light, soil moisture, temperature, etc.) and automating the irrigation system is installed in IoT-based smart farming. Farmers from almost anywhere can track the field conditions. When compared to the traditional method, IoT-based smart agriculture is highly effective.
- **Diagnostics**: The sensor data is inserted into a cloud-based IoT interface with pre-programmed decision rules and frameworks that decide the state of the entity being examined and recognize any defects or requirements. The sensor data is fed into a cloud-based IoT platform with pre-programmed decision rules and models that decide the state of the entity being examined and recognize any defects or requirements.
- **Decisions**: Good farming, in a way, involves making nuanced decisions based on the interrelationships of a multitude of variables, including crop requirements, soil conditions, extreme weather, and more. Traditionally, farming techniques have been assigned to a specific region or, at most, a part of one. Agricultural machine learning facilitates much more consistency, allowing farmers to interact almost directly with plants and animals, which in turn significantly enhances the effectiveness of farmers' decisions.
- **Action**: The loop starts again after the end-user assessment and intervention.

**Benefits of Smart Farming** In many ways, technologies like ML and the Internet of Things have the capacity to modify agriculture. There are five ways that ML and IoT can support agriculture:

1. Loads of data gathered by smart farming sensors related to climate conditions, soil fertility level, improvement in plant growth. or health of farm animals can be used to monitor the overall health of organization, as well as employee results, equipment quality, etc.
2. Internal processes are better regulated, and as a result, output risks are minimized. The ability to predict the success of production enables preparation for wider optimization of the commodity.
3. Pricing strategy and elimination of waste as a result of improved production control. The risks of wasting yield can be minimized upon detecting any irregularities in crop growth or livestock health.

**Fig. 4.1** ML and IoT-based
smart farming cycle

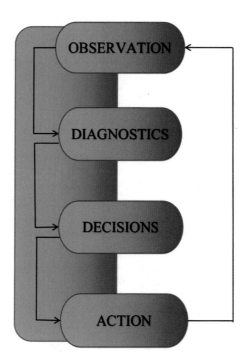

4. Digital transformation improves business productivity. Several processes can be
   automated throughout the development cycle with smart devices, such as irrigation, fertilizing, and pest management.
5. Quality of service and quantities are improved. By automating the production
   process, more leverage is there over the process to sustain greater crop quality
   and growth potential.

The remainder of this paper is as follows. Section "Literature Review" presents
a brief literature review of some of the research papers proposed and presented in
this area. Section "Applications of Machines Learning in a Major IoT-Based Smart
Farming Environment" discusses about the machine learning applications in big
IoT-based smart farming. Followed by that, in Section "IOT-AI-Cloud in Agriculture—The Need and Implementation," we present a combined framework for IoT, AI,
and cloud technologies with respect to disciplines in the agricultural sector. Finally,
we conclude our discussion in Sect. "Conclusion" presenting some of the emerging
research axes in this field.

## Literature Review

In this section, a brief review of some of the papers published in this field is presented.
Vji et al. [5] have proposed a tracking system whose prior aim was to figure out
excess irrigation, soil erosion, and crop-specific irrigation problems by developing

a spread wireless sensor network, in which every area of the field is surrounded by variety of sensors that transmit data to a central server. The pledge of IoT and wireless sensors in agricultural sector is illustrated by Ayaz et al. [6], along with the difficulties that will occur while combining this technique with conventional farming practices. Balducci et al. [7] concentrate on how to treat heterogeneous data and knowledge extracted from actual datasets which capture physical, biological, and sensory values. Dewangan [8] focused on the identification and determination of the standard and nature of the soil for a specific region, taking into account the level of toxicity at the present time and predicting its later usefulness using the AI model. Farooq et al. [9] have performed a structured literature survey of IoT technologies and their present use in various agricultural scenarios. The underlining characteristics of Gondchawar et al. [10] for smart farming includes a intelligent remote-controlled robot with GPS to carry out activities like spraying, scaring birds, sensing humidity, weeding and animals, maintaining vigilance, etc. Jha et al. [11] have addressed various automation activities such as IoT, wireless communications, and artificial intelligence (including ML and DL) and suggest a framework that can be implemented for identifying leaf and flowers and watering with the help of IoT in botanical farming. A detailed overview of research on ML applications in agricultural manufacturing systems was presented by Liakos et al. [12]. They also suggest innovative ideas for how ML and IoT can be integrated with such applications. Madushanki et al. [13] aim to review recently established IoT implementations in agricultural sector to furnish a summary of sensory data gatherings, techniques, and sub-verticals including crop and water management. McQueen et al. [14] narrate a project that applies a number of ML methods to agricultural and horticultural challenges. A brief survey is provided by Santos et al. [15] on various DL methods for different agricultural issues, like disease identification or detection, classification of plants, and fruit counting among other fields. Stoces et al. [16] gave a description of the categorization of IoT equipment, platforms, protocols, and network solutions. It concentrates on network infrastructure, which acts as the base for IoT deployment. Syed et al. [17] have proposed a smart irrigation system which would aid in suitable water management and provide perfect crop recommendations considering historic soil state data, as well as the number of minerals necessary to be included in the soil.

## Applications of Machines Learning in a Major IoT-Based Smart Farming Environment

ML is a useful way for devices to imitate human learning habits, acquire new knowledge, enhance output over time, and attain the desired maturity. It has been very influential in algorithms, hypotheses, and implementations over the past couple of years, together with other farming techniques to reduce crop prices and optimize yield. ML technologies on agricultural farms can be commonly used in domains such as pathogen detection, seed identification, irrigation planning, soil quality, weed identification, grain quality, and weather prediction. It may be used after harvesting

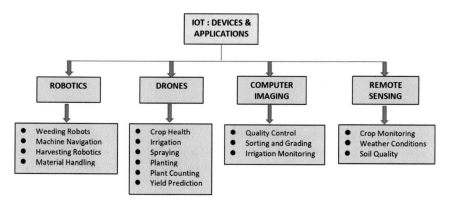

**Fig. 4.2** Taxonomy of IoT applications

to analyze the vitality of produce (fruit and vegetable freshness), shelf life, product consistency, competitive analysis, and so on. Support vector machines (SVMs), discriminant analysis, fuzzy clustering, K-nearest neighbor, naive Bayes, K-means clustering, Gaussian mixture models, decision making, artificial neural networks (ANNs), including deep learning could all be used in IoT-based farming [18]. ML has a massive impact on the efficacy of crop categorization and quality, development of agrochemicals, identification of diseases, treatment, and management. Figure 4.2 shows the taxonomy of some of the IoT applications in agriculture [19].

The applications of machine learning that have been involved in IoT-based smart farming are as follows:

1. **Diseases Management**: It is possible to use blended ML and IoT to identify and monitor diseases in agricultural areas. In order to protect crops from these pathogens and minimize labor, ML methods further stimulate suitable pesticides. Such a system helps farmers by gathering data and arranging fertilizers, pesticides, and irrigation appropriately. Fruits quality and quantity have been improved and extreme pesticide usage minimized by correctly recognizing the disease and giving accurate pesticide formulation and irrigation techniques. Besides that, architecture for defining and categorizing speech measures of various plants using deep learning techniques. At these plants, the audio steps focus on visual data captured in real time and travel through different regions of the farm through IoT-based camera sensor nodes deployed in agricultural areas.

2. **Soil Management**: Temperature, nitrogen levels, and moisture are only a few of the many soil characteristics that affect crop health. Farmers historically spread equal quantities per sq. meter of pesticides. Such a wasteful use of energy not only strongly impacts the budgeting of producers, but also flora and fauna tampers, lowering the amount of species of pollinators. The use of ML techniques in combination with wireless image processing tool can be used to determine the levels of soil erosion and the health of individual plants. The information collected is then used to assess the specific regions of the land which are infested, enabling farmers to use fertilizers precisely where they are required.

3. **Plant Management**: The crossover of ML and IoT offers a suitable and manageable atmosphere via greenhouse technology for agricultural plants. Traditional agriculture and environmental regulations, on the other hand, find it difficult to adjust to the growth of various types of plants at various stages of growth due to the spatiotemporal variability of crop growth environmental parameters and their reciprocal effects in safe agriculture. From a controlling and monitoring perspective, higher precision is therefore required. A variety of ongoing studies are being carried out to design and test forms of tracking and controlling systems for humidity and temperature adjustment, light, carbon dioxide concentration, and other climatic factors for IoT, technological, and economic performance. It is suggested that IoT, sensors, and actuators be used to monitor the environmental parameters for a particular type of plant.

4. **Weed Management**: For any farm, weed control is quite important. Autonomous flying machines to take pictures and map weed are proposed in a region to maximize this. An IoT network can be used to power a flying machine. NB-IoT and other advanced IoT techniques can manage and manipulate large quantities of data.

5. **Crop and Yield Management**: In farms based on collected data over the IoT environment, ML-based yield visualization could be applied via yield monitoring linked through GPS. The data obtained, which shows the yield information, will be mapped according to the different types of agricultural land. Aside from that, ML systems combined with IoT can be used to forecast and increase agricultural yields. These devices are used by farmers and others who have no previous programming experience. Crop out-come can be achieved with ML that produces data based on previously acquired information. This helps farmers to make decisions on agricultural crops that are economically sound.

6. **Water Management**: ML has been used to incorporate many systems for managing water supply for the farming sector as well as evaluating the quality of the water. It is possible to build IoT applications that use IoT sensors to detect surface parameters such as moisture content, water level, and climate changes. We can also use integrated ML and IoT systems to intelligently regulate the temperature of the water and adjust to the ambient temperature.

7. **Livestock Management**: To be economical and environmentally productive, livestock production systems require accurate data analysis and forecasting. ML can be used in different applications for livestock, including beef and milk processing, selective breeding, and more. It is very important to track animals in the field of agriculture. Many other studies have been carried out on animal monitoring using IoT-based sensors, and completely distinct studies have been carried out on the classification of animal types. IoT sensors may be used to confirm the location of an animal. ML methods may be used to identify and analyze tracked animals' living habits and body movement.

# IOT-AI-Cloud in Agriculture—The Need and Implementation

IoT applications in farming are targeted at traditional farming operations in order to satisfy rising demand and reduce production losses. IoT uses robots, drones, tracking devices, and computer imaging for crop monitoring, inspection, and crop scouting, as well as continually updating machine learning and analytical software, and provides farmers with data to save time and money on farm management. In this section, we will look at how IoT, ML, DL, and cloud technology work together to meet the needs of successful smart farming. Current agriculture systems are often unable to address the requirements of today's consumer, leading to a shortage of essential requirements such as processing speed, data storage potentiality, reliability, affordability, and flexibility. Even in computer-assisted agriculture systems, resources are underutilized. To solve the problem of current agriculture systems, it is important to establish a cloud-based autonomous information system that provides effective service [20]. Figure 4.3 shows the combined architecture with the assimilation of IoT, AI, and cloud systems. The services are categorized as software as a service (SaaS), platform as a service (PaaS), and infrastructure as a service (IaaS). Customers can communicate with the framework through a user interface in SaaS. PaaS offers flexible cloud middleware that enables cloud and user sub-systems to communicate. An autonomic resource manager can be used in IaaS to automatically pool and control resources. The given architecture comprises of three domains: (i) user domain, (ii) IoT domain, and (iii) cloud domain.

## *User Domain*

The user domain establishes a Web platform through which different classes of individuals interact with agents in order to supply and acquire useful agricultural data from respective domains. Climate, ground, seed, irrigation, profitability, machinery, disease, pesticide, and cattle are some of the distinct domains in farming that can be identified. Agricultural sector officers, farming specialists, and growers are the three groups of users that can be identified on the Web. The farming specialist will exchange professional expertise by resolving farmers' queries and upgrading the portal depending on the most recent findings in the area of farming in their domain. Agriculture inspectors are government representatives who are informed about the government's existing agricultural policies, systems, and rules. Growers are a critical community that stands to gain the most from posing questions and getting an automated reply after analysis. Users can monitor and receive responses to any data specific to their domain despite having to visit the farming support unit encouraging them to collaborate with a variety of farming domains. The user's requests are redirected to the cloud portal for latest update, and the feedback is delivered to them mostly through the Internet on their pre-configured gadgets (tablets, laptops, cell phones, etc.).

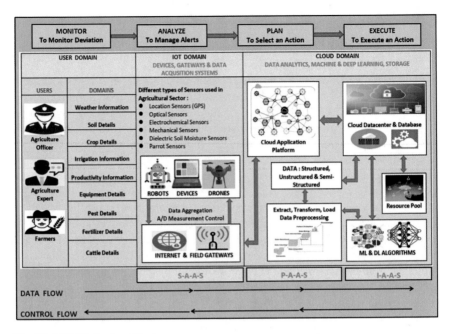

**Fig. 4.3** IOT-AI-cloud architecture

## IoT Domain

The Internet of Things (IoT) is an ecosystem of Internet-accessible, linked physical objects. In order to communicate with internal states or the external world, the embedded technology in the objects allows them, which in turn influences the decisions made. The IoT describes a system where, through wireless and wired Internet connections, objects in the real world and sensors inside or linked to these objects are hooked up to the Internet. These sensors can use different types of protocols for local area networking, such as RFID, NFC, Wi-Fi, Bluetooth, and Zigbee. Sensors may also provide protocols for broad-area communication, such as GSM, GPS, GPRS, 3G, and LTE. Table 4.1 lists some of the sensors and devices used in the agricultural sector along with their uses [15, 16].

## Cloud Domain

This domain is the site of the cloud-based platform for agriculture services. Users' information and a cloud infrastructure are used to preserve agricultural data in separate groups for distinct realms, each with its own unique identifier. The data is constantly tracked, analyzed, and processed. Selection, data preprocessing, transformation, categorization, classification, and interpretation using data mining, machine

**Table 4.1** Sensors used in agricultural sector

| Sensor | Usage |
|---|---|
| Location | Locates latitude, longitude, and altitude to just a few feet using GPS satellite signals |
| Optical | Used to determine the content of the soil's clay, organic matter, and moisture. They are normally connected to drones |
| Electro-chemical | Provides vital details about the pH and nutrient levels in the soil. |
| Mechanical | Used to assess the soil's mechanical resistances |
| Dielectric | Uses the dielectric constant to determine the moisture level of the soil |
| Parrot | Used to keep track of the temperature, moisture, and salinity of the soil around the plant. Data is sent to the farmers' mobile |
| Spruce | Used to monitor irrigation. Data is stored on a cloud server, which the user can access at any time, and from any place |
| Koubachi | It is used to sprinkle water on the plants. It serves as a node that gathers data from a variety of sensors, such as air temperature, soil moisture, and sunlight |

learning, and deep learning processes are all part of the research process. User data is classified in the storage repository based on various predefined domain classes. Via pre-configured computers, for final confirmation, this information is then passed to agriculture professionals and officers. At the infrastructure level, the autonomic system administrator determines resource needs automatically relying on scalability specifications and then allocates and implements resources. Data collected exists in the form of structured, unstructured, and semi-structured formats. The Knowledge Data Discovery (KDD) mechanism facilitates four sub-processes (i) data cleaning, (ii) data integration, (iii) data transformation, and (iv) data contraction. Data transformation serves as a link between the data analysis (classification) and data preprocessing sub-processes. After data preprocessing, this method converts the labeled data into a suitable format for classification. Agriculture data from different users from various domains is graded at the classification stage, which is dependent on the derived data. ANN, SVM, K-nearest neighbor (KNN), etc., are examples of machine learning algorithms which are used for regression and clustering purposes. In the agriculture field, as we saw in the previous section, ML is used in a wide variety of areas including crop management, livestock management, disease management, etc. Deep neural networks, such as convolution neural networks (CNN), are one of the most frequently used techniques for finding patterns in soil and plants in computer vision and classification in the agricultural sector.

Thus, all these domains work interconnected to each other by continuously monitoring the deviations in sensor working, analyzing, and managing the alerts, sustainability planning, and executing all the actions for smooth functioning.

## Smart Farming Challenges

Smart farming works in four layers: perception, network, edge, and application. Each layer has its own set of challenges giving a significant drawback. Perception layer works with physical devices such as sensors and actuators. Accidental or malicious human activity, viruses, malware, or cyber criminals may cause physical devices to malfunction. Smart farming systems use a wide range of sensors and technologies, which opens the door to a number of security risks. Network layer challenges mainly deal with network and denial of service issues. Smart farming necessitates an uninterrupted or continuous Internet link to be competitive. This means that this farming method is completely unworkable in rural communities, particularly in developing countries where mass agricultural production is common. Smart farming would be impossible in areas where Internet services are painfully slow. Smart farming, as previously mentioned, requires use of high-tech tools that necessitate technical ability and precision to be effective. It necessitates a working knowledge of robotics and information technology. Many farmers, however, lack these abilities. Finding anyone with this level of technical knowledge is challenging, if not impossible. Critical elements at the edge track and control sub-systems communicate with all levels and gain access to strategic resources. Instead of being centralized in the cloud, the processing of vast amounts of data produced by the perception layer can be done locally. This layer can provide services with a faster response time and higher quality due to the distributed architecture of edge computing. Smart farming systems are made up of a set of devices that communicate with one another and have varying degrees of limitations. Many flaws arise as a result of device limitations, which render it difficult to use current tools and protection techniques. To ensure effective and secure operations, top-layer appliances with powerful computing capabilities may implement more robust security mechanisms. Figure 4.4 shows the various types of challenges probable in smart farming. Figure 4.4 shows the various types of challenges probable in smart farming.

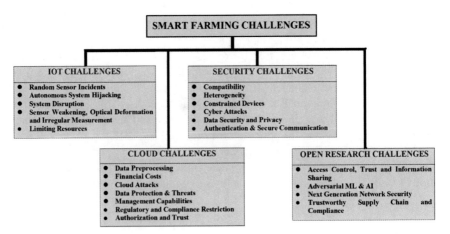

**Fig. 4.4** Smart farming challenges

# Conclusion

Smart farming has the capability to provide a more productive and feasible form of agriculture sector by relying on a more systematic and resource-efficient approach. New farms will bring humanity's long-held dream to fruition in the near future. Agriculture driven by IoT, ML, and cloud is the first step in the development of smart farming. With the aid of the agricultural IoT, ML algorithms have been developed to data produced from different farm inputs to make the model intelligent and to provide conclusive information and forecasts. Current applications of ML in agriculture are examined in this research, from method to outcomes. By introducing ML to sensor data, farm management systems evolve into a real artificial intelligence (AI) system that provides deeper recommendations and observations for subsequent job decisions and behavior with a variety of finished product enhancements. This selection will be expected in the future and will make broader use of ML models possible. In the end, the engagement of cloud-based platforms with ML applications in IoT-based smart farming is also discussed. In future, a Web-based system can be implemented using IoT and DL to detect diseases in the crop. This system will take the user's input image and produce an output in the form of disease detection, preventive measures, pesticides needed, and the likely cost of recommended pesticides.

# References

1. Abhishek, L., Barath, R.: B.: Automation in agriculture using IOT and machine learning. Int. J. Innov. Technolo. Exp. Eng.(IJITEE). **8**(8), 2279–3075 (2019)
2. Doshi, J., Patel, T., Bharti, S.K.: Smart Farming using IoT, a solution for optimally monitoring farming conditions. Procedia Computer Sci. **160**, 746–751 (2019)
3. Mulge, M., Sharnappa, M., Sultanpure. A., Sajjan. D., Kamani, M.: Agricultural crop recommendation system using IoT and M.L. Int. J. Anal. Experim. Modal Anal. **12**(6), 1112–1117 (2020)
4. Bhanu, K. N., Jasmine, H. J., Mahadevaswamy, H. S. : Machine learning implementation in IoT based intelligent system for agriculture. In: 2020 International Conference for Emerging Technology (INCET), pp.1–5. IEEE (2020)
5. Vij, A., Vijendra, S., Jain, A., Bajaj, S., Bassi, A., Sharma, A.: IoT and machine learning approaches for automation of farm irrigation system. Procedia Computer Sci. **167**, 1250–1257 (2020)
6. Ayaz, M., Ammad-Uddin, M., Sharif, Z., Mansour, A., Aggoune, E.H.M.: Machine learning applications on agricultural datasets for smart farm enhancement. IEEE Access **7**, 129551–129583 (2019)
7. Balducci, F., Impedovo, D., Pirlo, G.: Internet-of-Things (IoT)-based smart agriculture: toward making the fields talk. Machines **6**(3), 38 (2018)
8. Dewangan, A.K. : MApplication of IoT and machine learning in agriculture. Int. J. Eng. Res. Technol. (IJERT), **9**(7) (2020)
9. Farooq, M.S., Riaz, S., Abid, A., Umer, T., Zikria, Y.B.: Role of IoT technology in agriculture: a systematic literature review. Electronics **9**(2), 319 (2020)
10. Gondchawar, N., Kawitkar, R.S.: IoT based smart agriculture. Int. J. Adv. Res. Computer Commun. Eng. **5**(6), 838–842 (2016)

11. Jha, K., Doshi, A., Patel, P., Shah, M.: A comprehensive review on automation in agriculture using artificial intelligence. Artif. Intell. Agric. **2**, 1–12 (2019)
12. Liakos, K.G., Busato, P., Moshou, D., Pearson, S., Bochtis, D.: Machine learning in agriculture: a review. Sensors **18**(8), 2674 (2018)
13. Madushanki, A.R., Halgamuge, M.N., Wirasagoda, W.S., Syed, A.: Adoption of the internet of things (IoT) in agriculture and smart farming towards urban greening: a review. Int. J. Adv. Computer Sci. Appl. (IJACSA) **10**(4) (2019)
14. McQueen, R.J., Garner, S.R., Nevill-Manning, C.G., Witten, I.H.: Applying machine learning to agricultural data. Comput. Electron. Agric. **12**(4), 275–293 (1995)
15. Santos, L., Santos, F.N., Oliveira, P.M., Shinde, P.: Deep learning applications in agriculture: a short review. In: Iberian Robotics Conference, pp. 139—151. Springer, Cham (2019)
16. Stoces, M., Vanek, J., Masner, J., Pavlikk, J.: Internet of things (IoT) in agriculture-selected aspects. Agris Online Papers Econ. Inform. **8**(665-2016-45107), 83–88 (2016)
17. Syed, F. K., Paul, A., Kumar, A., Cherukuri, J.: Low-cost IoT+ML design for smart farming with multiple applications. In: 2019 10th International Conference on Computing, Communication and Networking Technologies (ICCCNT), pp. 1–5. IEEE (2019)
18. Maduranga, M.W.P., Abeysekera, R.: Machine learning applications in IoT based agriculture and smart farming: a review. Int. J. Eng. Appl. Sci. Technol. **24** (2020)
19. Kamilaris, A., Prenafeta-Boldú, F.X.: Deep learning in agriculture: a survey. Comput. Electron. Agric. **147**, 70–90 (2018)
20. Gill, S.S., Chana, I., Buyya, R.: IoT based agriculture as a cloud and big data service: the beginning of digital India. J. Organ. End User Comput. (JOEUC) **29**(4), 1–23 (2017)
21. Naresh, M., Munaswamy, P.: Smart agriculture system using IOT technology. Int. J. Recent Technol. Eng. **7**(5), 98–102 (2019)
22. Sowmiya, M., Prabavathi, S.: Smart agriculture using IoT and cloud computing. Int. J. Recent Technol. Eng. **7**(6S3), 251–255 (2019)

# Chapter 5
# Smart Airport System to Counter COVID-19 and Future Sustainability

**Uma N. Dulhare and Shaik Rasool**

## Introduction

The latest threat that changed the entire world is coronavirus (COVID-19) that impacted every aspect of the society. It has thrown us with several challenges as well as opened new opportunities to explore ways of living life. It has caused severe damage to health, wealth, and living if individual and communities/we have seen people losing jobs, companies losing wealth and sadly several deaths that shock the entire world. People have been confined to homes and became mentally stressed and emotionally weak due to lockdowns imposed by the governments. Several lost their livelihood and became financially ill. Every pandemic provides us an opportunity to obtain critical information that may lead us to live sustainable life and may help us to avoid future events. It may not be fully possible to avoid such in future completely, but it will train us to reduce the impact or make us ready to handle such crisis.

COVID-19 has awakened us and has presented challenges to prepare ourselves with a need of comprehensive study in natural sciences of sapiens and exposure constraints to them. This requires continuous surveillance, research, and apt diagnosis. Figure 5.1 shows a graphical representation of the current cases around the world as on 11 Feb 2021. As we are aware that prevention is the best strategy that can be applied in current situation to contain further spread of the virus in society. As there is no approved treatment available for COVID-19, early detection, containment, and rapid preventive measures are only means to stop the spread. Preventive measures should include careful drafting of steps for early detection, isolation, and creating awareness among the people in society by educating them with things that should be adapted and that may be avoided [1].

U. N. Dulhare (✉) · S. Rasool
Muffakham Jah College of Engineering, Hyderabad, Telangana, India

Methodist College of Engineering and Technology, Hyderabad, Telangana, India

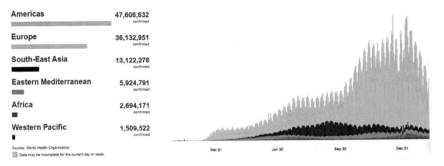

**Fig. 5.1** Situation by WHO region. Globally, as of 5:01 pm CET, 11 February 2021, there have been 106,991,090 confirmed cases of COVID-19, including 2,347,015 deaths, reported to WHO [2]

While the loss has spread to every industry and sectors in national and international levels, it is quite clear that airports and railways stand as the biggest threat and centre for spreading of virus across the borders. A chapter on "Innovative Applications of Big Data in the Railway Industry" helps to solve most of problems in the railway sector [3]. Aviation industry is in a deep state of distress over the current pandemic situation. Fear for health and security concerns has grown rapidly and made individuals to confine to their homes worldwide. Not only the passengers but airline staff are concerned and stressed. It shows that the crisis may exist longer and may require a shift in the strategically directions that need to be incorporated currently to overcome this pandemic crisis in aviation industry. Technologies infused with ease of user experience is required to strengthen the confidence levels of all. Risk management strategies should be developed to counter the risk with proper planning and scheduling of activities. Passenger and travel data will play a vital role in designing the countermeasures. Data like the passenger health status, travel experience, past health records, travel time, etc., are very crucial in planning.

The proposed model will make it possible to calculate the number of passengers to be processed in accordance with the available check-in counters based on the proposed measures. It will also assure that future pandemic events could be avoided, and the risk of economic and asset loss could be greatly minimized. The proposed model will assure safety and health keeping in mind the necessity and need of current and future. It incorporates latest technologies like blockchain, machine learning, and IoT to gain the confidence back by assisting in maintenance and diagnosis of the passengers. Privacy of passengers and security are considered and carefully handled to avoid any future concerns. Safe travel and health are need of the hour and are the central core for this proposed model.

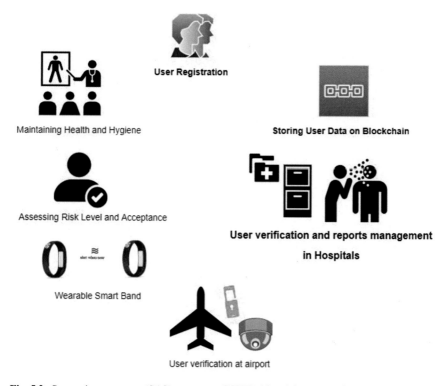

**Fig. 5.2**   Smart airport system (SAS) to counter COVID-19 and future sustainability

## Proposed System

This chapter proposes a model to consider the effects and severity of pandemic and provide a model to safeguard the airport which forms the central and critical point of disease spread. To prepare ourselves, using the technological innovation in this era is a challenging task. Technology has brought us ease of use as well as concerns like security, privacy, theft, etc. Our model is designed keeping in view all necessities and concerns of every individual as can be depicted in Fig. 5.2.

## Passenger Registration Using Blockchain

The most vital part of any system is the user data. The entire passenger who holds a passport has to be registered, and their data should be stored for screening purpose. The data will contain their complete medical history and identification information including their facial data. There are several solutions available to handle the user data. Cloud services nay look apt for this sole purpose of storing and handling the data. They have been proven to provide reliable services over the stable Internet. But they also come with challenges like bandwidth, efficiency, and privacy concerns which are very crucial. Several attacks on various cloud services in the recent years have supported our concern that it may be infeasible for our model. Data leaks prove that a decentralized model like blockchain is pretty much the need of today era eliminating the risk and fear of doing business in all sectors. Slowly its growing pace and trust prove its effectiveness.

## Blockchain for Data Storage

To ensure the trust of customers and to achieve security, blockchain stands as a very viable solution. There is decentralization in every transaction that is segmented and stored over the grid [4]. It withstands the outside attempts that are made to access the data with a ledger adapted in place. Blockchain also ensures capacity and availability of data everywhere with decentralization. Confidentiality, authentication, and integrity are the pillars of security which are well handled and provided by blockchain. It diminishes the need of monitoring the transactions done and sharing by providing proof of authentication, thereby increasing the overall confidence in the entire system.

## Blockchain Data Storage

The process of storage begins with processing of data and segmentation. Then the segments are distributed over the decentralized system as shown in Fig. 5.3. The process can be understood in the below steps:

- Generating shard data

Shard are created by segmenting the data that is to be stored on the blockchain.

- Encryption of shards

Shards cannot be stored as it is due to concern for security. Encryption is used to provide the needed confidentiality measure for storing the shards. Shards are first encrypted then stored.

- Calculation of Hash

For each shard, a hash is generated that ensures the integrity of shards. A special hash is generated for each of these shards.

- Shard replication process

Replication of shards is done to obtain the redundant copies to ensure the availability.

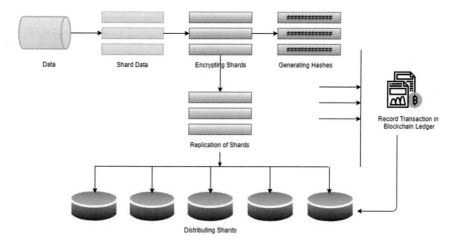

**Fig. 5.3** A blockchain storage system

- Shard distribution in decentralized system

To ensure the availability of shard in the blockchain the replicated shards are distributed across the system.

- Recording the transactions

A ledger is maintained to store and records all these transactions which forms the central part of decentralized blockchain system.

All the transaction in the blockchain is synchronized across all nodes in the system. The data to be stored in blockchain are customized and optimized into a form of transaction before storing to ensure the effectiveness of storage. All the transactions stored in blockchain forma secure time stamped logs of data. This ensures that the data stored in blockchain is immutable and extremely secure. Also, there is transparency maintained as public has access to it [5]. Access to certain attributes like location and customization may not be permitted in certain cases. In such case, addresses are used to store the data in form of small bits as part of blockchain technology. Thus, secure data storage is possible when addresses are forwarded as transaction after encoding them. The storage location is transformed automatically into the address of the receiving system to ensure the security of data transmission.

## *Benefits of Storing Data Using Blockchain*

- Cloud storage is expensive than blockchain.
- No specific equipment is essential.
- Managing and controlling is easier and eliminates the need of an administration.
- It is more transparent than traditional approaches.

- The system offers immutable, authenticated, and secure transactions.
- It ensures availability with replication across distributed decentralized system.

## Passenger Verification and Reports Management in Hospitals

Future of blockchain extends a perfect solution for medical data security. It is proven that blockchain technology is indeed a perfect solution for various medical data security issues even at global level. As all the passenger data are stored using blockchain, whenever a passenger visits hospital for any kind of treatment or diagnosis, their verification is done using facial recognition system. Once the passenger is verified, then their records are retrieved from blockchain for validation and updating. If the passenger is to travel, then a passenger must visit any hospital and get a check-up. Upon check-up, the hospital may mark the record as safe. If any ailments are found, then deep learning techniques are employed to analyse the patient medical history and make recommendation for treatment or mark them as unsafe as shown in Fig. 5.4.

## Passenger Authentication at Airport

When the passengers arrive at the airport, their luggage is screened and UV sanitized. They must be authenticated using facial recognition system and check to see the status and recommendations from the hospital. The challenge lies in how facial recognition would be done on a passenger wearing a mask; first the passenger must be verified that he is wearing a mask, and second, person must be authenticated using AI with blockchain data as shown in Fig. 5.5.

Hanwang's claims that its cameras are now equipped with AI techniques that could recognize the faces with masks with an accuracy of 95%. To achieve this, it captures and processes multiple photos of the target to aptly recognize the face in real time. A study conducted at University of Bradford claims that the software designed by them can recognize the faces by deleting the half image containing the mask in the target. Only the features available in the half image are used to recognize the facial points and provide accurate results. It is prompt guessing of faces that may look like when they are wearing masks as with images that are available in the database. To attain accuracy, a system would first require training the system by using images of the targets with face masks as shown in Fig. 5.6.

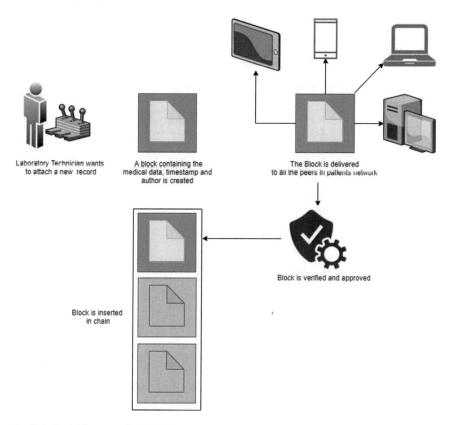

**Fig. 5.4** Model for managing EHR's

**Fig. 5.5** Surveillance camera identifying faces with mask

## Recognizing Faces with CNN

## Dataset of Facial Images

First step involves building a dataset of images of persons with mask and without masks. We may require multiple images of persons to achieve accuracy. All the

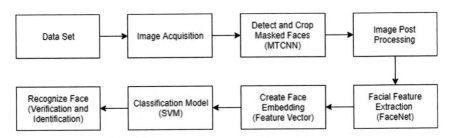

**Fig. 5.6** Process of facial recognition with mask

passengers registering in the system should undergo this step to ensure availability and accuracy.

## MTCNN System for Face Detection

Detecting the facial part in an image is crucial. A model multi-task cascaded convolutional neural network (MTCNN) proposed by Zhang, Li, Zhang, and Qiao accurately performs the task of detecting the faces from an image and is more efficient when compared to other techniques. Thus, a pretrained MTCNN model may be used to detect the facial part that may be masked. We obtain a set of facial descriptors from the images. Three surging networks are part of this model. At first, an image pyramid is built by scaling the obtain image sets. Then, facial regions are marked in the images utilizing the first proposal network. The bounding boxes are refined in the later stage utilizing refine network. As a last step, output network obtains the facial landmarks. The output is subjected to maximum suppression to refine bounding boxes [6].

## Post-Processing of Images

Faces are cropped and resized as needed from the input image by utilizing the bounding box obtained in MTCNN. The apt resized image should be $160 \times 160$ pixels as recommended in FaceNet architecture.

## Facial Feature Extraction

FaceNet is encapsulation of deep CNN and batch layer. Deep CNN is validated by 1.2 normalization. Face embedding is obtained as outcome of normalization. FaceNet is developed on 22 deep CNN and 128-dimensional embedding. The output is trained on these layers to utilize it as face descriptor. These work as similarity comparative system using embedding module. Max operator is applied on the feature set to acquire a unique feature vector. The network is fine-tuned to work on heterogeneous tasks of facial verification.

## *Facial Authentication*

To recognize the faces, the system employs classification using support vector machine (SVM) as the last step in recognition process. It produces robust output to overfitting. Class separation efficiency is the margin. We compare the target face with that of trained face data set using SVM. The minimum distance obtained is considered as correct identification of face. Normalization method is used for similarities checking.

## Wearable Smart Band

Wearable technologies are smart devices that are in touch with your skin. These wearable devices having sensors can analyse user physical and mental health. User can get notified of every activity on the display of the wearables as shown in Fig. 5.7. You can receive calls, SMS and able to see and receive calls by connecting smart-watches with mobiles. Fitness bands are the health monitors. These bands show heart rate, calories burned, step walked, blood pressure, time spend on exercise, total steps in a day, health risk assessment, and many more. These wearable devices are worn during exercise and daily use [7].

Wearable technology in health care can monitor physiological and biomedical parameters. These parameters include blood pressure, body temperature, oxygen saturation in the blood, heart rate, electrocardiogram (ECG), and ballistocardiogram (BCG). The advanced features of wearable technology in the healthcare industry will help to improve the health of the patients. It will reduce the cost of care. Data

**Fig. 5.7** Social distancing alert system using smart band

generated in the wearable device are helpful to analyse the health of the user quarterly. With this technology, it is easy to get information about deficiencies in the body [8, 9].

At the airports, these wearable bands are used to analyse the current condition of the patient like his body temperature, ECG, etc. They are also used to alert a passenger whenever he comes nearest to another passenger as shown in Fig. 5.7 to maintain social distance. They are given to the passenger on arrival, and passenger needs to wear it till completion of his travel to destination [10, 11].

The wearable band may also be used to continuously monitor the patient throughout the travel time. Emergency can be avoided, and necessary steps can be taken before situation becoming worse.

## Assessing Risk Level and Acceptance

Once the passenger verification has been done using blockchain and facial recognition, the data generated from wearable device are used to analyse the current health condition of the passenger at risk assistance and acceptance counters as shown in Fig. 5.8. Support vector machine (SVM) algorithm can be used to assess and classify the amount of risk associated with the passenger. If the risk level is low or none, then the passenger may be allowed to travel and issued a boarding pass. If the risk is moderate, then necessary steps may be taken to minimize the risk such as screening at airport to diagnose the exact problem and provide medical support. Based on the problem identified, the passenger may be allowed to travel with necessary precautions or may be sent back. If the level is high, they may be immediately shifted to hospital, and their relatives may be informed.

**Fig. 5.8** Risk assistance and acceptance counters

## Maintaining Health and Hygiene

A small investment could be made by airports in installation and maintenance of smart infrastructure like contact free spouts and sanitizer dispensers in the restroom, AI-enabled robotic cleaners. Efficiency can be improved by adopting usage based cleaning and avoiding scheduled mainteannce which will also result in safe travel experience. Restroom occupancy by the passengers can be monitored and maintained by installing AI-based solutions in real time that stipulates usage patterns and triggers maintenance alerts to sanitation staff for freshly inhabited sections [12]. Airports may maintain separate rooms for passengers who are ill and may also be used to quarantine the employees when necessary. Then, prompt medical treatment can be given for quick recovery and avoiding further spread.

### Maintaining Clean and Hygiene Places

An IoT device constituting an ambient temperature and humidity sensor can be set up at the airport rest rooms or longue to keep the environment inside stable. The flow of people in the rest room can be a measure for cleaning work to be carried out. A challenge here is that people may be men, women and children of different age groups and size. Setting a certain threshold will help us to make apt decision keeping clean and hygiene places. The designated staff will carry out the cleaning work when they receive alert after threshold value is met.

### Smart Soap Dispensers

Smart soap dispensers are very useful to passengers as they dispense without requiring physical touch to the dispenser. The right amount of soap is dispensed when needed. It can also be used to keep track of number of people using the service. Alerts can be generated when refill is required avoiding continuous monitoring of soap levels in the dispenser.

### Smart Toilet Paper Dispenser

A smart toilet paper dispenser reduces the amount of time monitoring the level of paper left like as in smart soap dispenser discussed above. It may also have an UV lamp to disinfect the left-over paper in the dispenser to give confidence of clean and hygiene papers. The entire rest room may also be exposed to UV sanitization to ensure it free from viruses in every use. The occupancy of rest rooms can also be monitored using application which will show the status of rest room occupied by others.

### Smart Garbage Bin

Smart garbage bin will be equipped with UV sanitization to ensure any viruses in the waste are killed before being handled by cleaning staff. The level of garbage bin can also be monitored through apps and bin alerts when it is full and need to be emptied. It will also be opened automatically for collecting waste, thereby avoiding any need to touch the surface. A futuristic model may also automatically dispose the waste into garbage collecting container using AI.

## Conclusion

Airports today need to be well equipped to handle large volumes and be prepared for any future pandemic events by identifying the patterns that are of concern. It may not be an effective approach if implementation is done by a few airports, rather it needs to be strictly enforced in all airports for a sustainable future. It is critical for all airports to work collaboratively and share information with all for achieving a success in solution built on technological framework, thereby resulting in desired results. Building and deploying such smart technological solutions may not be easy in current crisis and state of industry, but it may be right time to implement fruitful changes in view of low passenger travel and future insights. Once the deployment begins, there will be a need to educate the staff and the passengers to have cooperation and collaboration between them to yield transformation success. Faith and support of the authorities and management are also very crucial for technological transformation to gain the customer confidence and restore a safe and hygienic travel experience to all.

## References

1. Güner, R., Hasanoğlu, I., Aktaş, F.: COVID-19: prevention and control measures in community. Turkish J. Med. Sci. **50**(SI-1), 571–577 (2020). https://doi.org/10.3906/sag-2004-146
2. World Health Organization. WHO Coronavirus disease (COVID-19) dashboard (2021). https://covid19.who.int/
3. Shaik, R., Dulhare, U.N.: Evolution of Indian railways through IoT. In: Kohli, S., Senthil Kumar, A.V., Easton, J.M., Roberts, C. (eds.) Innovative Applications of Big Data in the Railway Industry, pp. 269–290. IGI Global, Hershey, PA (2018). https://doi.org/10.4018/978-1-5225-3176-0.ch012

4. Esposito, C., De Santis, A., Tortora, G., Chang, H., Choo, K.R.: Blockchain: a panacea for healthcare cloud-based data security and privacy? IEEE Cloud. Comput. **5**(1), 31–37 (2018). https://doi.org/10.1109/MCC.2018.011791712

5. Team Writer. Blockchain storage: meet your storage needs. Tech Funnel, 13 May 2020. https://www.techfunnel.com/information-technology/blockchain-storage/

6. Ejaz, M.S., Islam, M.R.: Masked face recognition using convolutional neural network. In: 2019 International Conference on Sustainable Technologies for Industry 4.0 (STI). Dhaka, Bangladesh, pp. 1–6 (2019). https://doi.org/10.1109/STI47673.2019.9068044

7. John Dian, F., Vahidnia, R., Rahmati, A.: Wearables and the Internet of Things (IoT), applications, opportunities, and challenges: a survey. IEEE Access **8**, 69200–69211 (2020). https://doi.org/10.1109/ACCESS.2020.2986329

8. Waliba, M.A., Ashour, A.S., Ghannam, R.. Prediction of harvestable energy for self-powered wearable healthcare devices: filling a gap. IEEE Access **8**, 170336–170354 (2020). https://doi.org/10.1109/ACCESS.2020.3024167

9. Cinel, G., Tarim, E.A., Tekin, H.C.: Wearable respiratory rate sensor technology for diagnosis of sleep apnea. In: 2020 Medical Technologies Congress (TIPTEKNO). Antalya, Turkey, pp. 1–4 (2020). https://doi.org/10.1109/TIPTEKNO50054.2020.9299255

10. Desai, K., Mane, P., Dsilva, M., Zare, A., Shingala, P., Ambawade, D.: A novel machine learning based wearable belt for fall detection. In: 2020 IEEE International Conference on Computing, Power and Communication Technologies (GUCON). Greater Noida, India, pp. 502–505 (2020). https://doi.org/10.1109/GUCON48875.2020.9231114.

11. Vijayan, V., McKelvey, N., Condell, J., Gardiner, P., Connolly, J.: Implementing pattern recognition and matching techniques to automatically detect standardized functional tests from wearable technology. In: 2020 31st Irish Signals and Systems Conference (ISSC). Letterkenny, Ireland, pp. 1–6 (2020). https://doi.org/10.1109/ISSC49989.2020.9180174

12. Sales Mendes, A., Jiménez-Bravo, D.M., Navarro-Cáceres, M., Reis Quietinho Leithardt, V., Villarrubia González, G.: Multi-agent approach using LoRaWAN devices: an airport case study. Electronics **9**(9), 1430 (2020). https://doi.org/10.3390/electronics9091430

# Chapter 6
# Early Prediction of COVID-19 Using Modified Convolutional Neural Networks

**Asadi Srinivasulu, Tarkeshwar Barua, Umesh Neelakantan, and Srinivas Nowduri**

## Introduction

Corona is a group of virus which originates in mammals, as a disease. Corona VIrus Disease (COVID-19) belongs to corona family, introduced by world health organization (WHO) on Dec. 31st 2019. The very first case was found in Wuhan city, China (Guandong state), known as Wu Flue. COVID-19 has longest genome of ribonucleic acid (RNA) approx. 26–32 k long. COVID-19 [1, 2] malignant might be the most widely recognized kind of disease among inquest folks in the entire world.

Most recent COVID-19 data in 2019 indicates a third significant reason for biting the dust from corona cases in USA, with around 161,460 cases, with 19.27% of new disease occurrences and 26,730 fatalities, indicating 8.34% of most malignant growth passing. Despite of the existing work with the fact that COVID-19 [3, 4] might be far reaching sort of danger, but due to its recognition in the early stages, it made good achievement rates and become exorbitant due to block movement of the condition. Accordingly, researchers are compelled to monitor and conclude its fundamentals to the expanded patient's endurances.

### Graph for Mortality

Machine learning (ML) is a subdivision of artificial intelligence (AI) [5, 6], which uses data to enable machines to learn to perform tasks on their own, which also help classifying and forecasting information from the given data set. On the other hand, statistical consistency refers to specific property of a statistical procedure such as the estimates of population parameters, confidential interval estimation and tests of

A. Srinivasulu (✉) · T. Barua
Data Analytics Research Laboratory, SVEC, BlueCrest University College, Monrovia, Liberia

U. Neelakantan · S. Nowduri
Department of Computer Science, PCC, 900 W Orman Ave, Pueblo, CO 81004, USA
e-mail: csrinivas.nowduri@pueblocc.edu

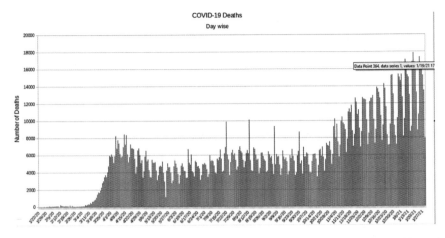

**Fig. 6.1** Real-time data set of COVID-19 worldwide. *Source* https://data.humdata.org/dataset/
novel-coronavirus-2019-ncov-cases

hypothesis and is a fundamental notation for supervised and unsupervised learning. Multi-class classification is ML [2, 7] task which consists of more than two classes or outputs. This research work focuses on a unified framework in studying the consistency of a general multi-class learning problem that concludes, by generalizing many known past results for specific learning problems (Fig. 6.1).

Majority of multi-class learning problems use an evaluation matrix based on a loss matrix, and as a result, algorithms for such problems are surrogate minimizing algorithms, which are characterized by surrogate [8, 9] loss. If surrogate loss is convex, that results in surrogate minimizing algorithm, it can be framed as convex optimization problem and can be solved efficiently.

This research study focuses on three directions. First part attempts to describe calibrated surrogates losses which leads to a consistent surrogated minimizing algorithm for a given loss matrix. It also discusses the necessary and sufficient conditions under which calibration will happen based on geometric properties of the surrogate and true loss. Second part focuses on discussing about convex calibration dimension that characterize the intrinsic difficulty, while achieving consistency for a training problem. Finally, we analyze the generic procedure to conduct convex calibrated surrogate.

In health data analytics field, computer vision (CV) helped acknowledgment based on conclusion convolution added design (CAD), which is truly blended with imaging trademark anatomist. On the other hand, the ML grouping demonstrates the status of planning and supports radiologists for exact analysis, diminishing time parameter and associate cost effectiveness in such determination.

The DL strategies indicate promising outcomes in an assortment of CV undertakings, such as division, arrangement and article discovery. These techniques comprise of convolutional [10] layers that can extricate various low-level nearby highlights to significant level worldwide. An associated layer toward end of the convolutional neural layers changes over tangled highlights into probabilities of specific names. For example, clump standardization layer (CSL) standardizes the contribution of a layer with a zero mean and a unit variation and dropout layer (DL), which is one of regularization strategies that overlooks the haphazardly chosen hubs. DL is also expected to improve the exhibition of profound learning-based techniques. Concluding focus is based on principle difficulties of profound learning-based strategies, which are applied to various fields, such as health clinical imaging and image database applications.

## Literature Review

According to the past research in this area, an immense amount of work has been done by people working at hospitals, clinics and laboratories, along with many researchers and scientists, dedicate considerable efforts in fight against COVID-19 epidemic [4]. Due to unconscionable dissemination of the disease, the implementation of AI made a significant contribution to the digital health district by applying the basics of automatic speech recognition (ASR) and DL algorithms. This study also focuses on the importance of speech signal processing, through early screening and diagnosing the COVID-19 virus by utilizing the convolutional [11] neural network (CNN). Particularly existing work, via its architecture, long short-term memory (LSTM) for analyzing the patient's symptoms such as cough, breath and voice. Our results find low accuracy in data set tests compared to coughing and breathing sound samples. Our results are in preliminary stage, possibly expected to enhance accuracy of the voice tests, by expanding the data set and targeting a larger group of healthy and infected people.

Recently, NAACCR announce that all malignancy cases are characterized by ICD for oncology with the exception of adolescence and pre-adult COVID-19, which were arranged by the ICCC. The root causes of deaths were grouped by the ICD. Whenever COVID-19 is attacked, disease frequency rates introduced in this report were balanced for delays in detailing, which happens on account of a slack in the event that catch or information redressal.

Past several researchers reported the automated screening and diagnosing, based on the analysis of chest CT images [1, 2, 7, 8]. AI is found to be clenched and enforced in e-health districts to aid early detection of COVID-19, by analyzing sound through coughing, breathing and speech [3]. The respiratory [12] sound is an indication for human health status which can be recognized and diagnosed by implementing ML algorithms [9].

Ever since the outbreak of COVID-19 virus, many scientists and researchers started considering the detection of COVID-19 from respiratory [12] sounds [13]. In similar studies, a low power consumed wearable system is proposed for detecting asthma and wheezing. This analysis is based on patient's frequencies of both sound features and respiratory [12] sound [14, 15]. In another significant study, a convolutional neural network (CNN) [16] is used in detecting different types of coughs, based on the analysis of their extracted sound features [17]. Besides, several systems are proposed for predicting COVID-19, using DL algorithms via classifiers such as CNN, long short-term memory (LSTM) and artificial neural network (ANN) [18]. Therefore, COVID-19 patients' health status can be determined through their speech signals. A patient's health state detection system can be used to observe and analyze the sleep-quality, severity of illness, fatigue and anxiety [19]. Since cough has been a symptom of many diseases, it is possible to distinguish between coughs, to establish the type of illness by testing the auditory features using multiple classifiers [20]. Moreover, in [21], a proposed system composed of a novel multi-pronged mediator AI architecture model is set to differentiate between the different types of coughs.

The latest year for which rate of mortality information is accessible slacks from 2 to 4 years behind the present year because of the time required for information assortment, accumulation, quality control and spread. The quantity of intrusive disease cases was assessed utilizing a 3-advance spatio-worldly model dependent on excellent rate of information from 50 states and the District of Columbia speaking to around 94% populace inclusion (information was missing for the entire years for Minnesota and for certain years for different states). This strategy cannot appraise quantities of basal cell or squamous cell skin malignancies since information on the event of these COVID-19 is not required to be accounted for to disease vaults.

Fig. 6.2 explains the statistics of the male versus female COVID-19 infection rates.

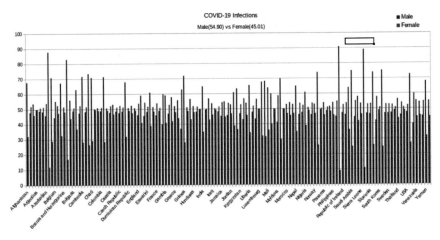

**Fig. 6.2** Male versus female COVID-19 infection rates. *Source* https://data.humdata.org/dataset/novel-coronavirus-2019-ncov-cases

## Existing System and Their Implementation Details

Currently, there are some existing systems for assessing the precondition of COVID-19 patients based on few NN techniques [15, 17]. All these used techniques and algorithms are found to be limited in terms of their performance accuracy and time complexity. In order to address these parameters, we propose MCNN algorithms as described in this work. This section in particular addresses the existing systems along with their drawbacks and then proposes a new systems as shown below.

### *Existing System*

In most of the existing systems, one can understood from the confusion matrices and the tables that the concatenated network performs better in detecting COVID-19 and not detecting false cases of COVID-19. That definitely outputs a better performance and overall accuracy [14, 15]. Their existing work clearly indicates that they had an imbalanced data set and a few cases of COVID-19, used in their proposed techniques, and does not yielding to the required results. In this work, we have improved COVID-19 early detection along with the other classes detection (such as positive or negative), through MCNN. We also identified the main reasons behind low precision in the past work is due to their data mining techniques, which we have eradicated in our research work with the help of deep learning [10, 22] techniques.

In another study, the results are presented in two different forms, such as 2 and 3 classes due to imbalance in the COVID-19 image data set, coupled with several meaningless results [22]. This experiment presented the results for each class and for all the classes with meaningful results that are more practical. We could have tested our network on a few cases like some of the other researchers have done recently, but we wanted to show the real performance of our network with few COVID-19 cases. As mentioned, mistakenly detecting 67 cases from 11,273 cases to be infected with COVID-19 is not very much but not very well also, and we hope that by using much-provided data from patients infected, COVID-19, the detection accuracy will rise much more. Convolutional [11] neural networks [4] were created because there were a few issues in the feed-forward neural network:

- Cannot handle sequential data.
- Considers only the current input.
- Cannot memorize previous inputs.

The solution to these issues is the CNN. ACNN can handle sequential data, accepting the current input data and previously received inputs. CNNs can memorize previous inputs due to their internal memory.

### Drawbacks of Existing System

- Less accuracy.

- High time complexity.
- Less performance.
- High computational cost.
- Uses more training COVID-19 image data set.

## *Proposed System*

Modified CNNs are the most mainstream profound learning models for handling multidimensional cluster information, for example, shading pictures. A run of the mill modified CNN [22] comprises of different convolutional [11] and pooling layers followed by a couple of completely associated layers to all the while get familiar with an element order and characterize pictures. It utilizes the combined model of back spread, a proficient type of inclination plunge to refresh the loads an interfacing its contributions to the yields through its multi-layered architecture. In this chapter, we present a two-phase approach utilizing two separate modified CNNs. The main modified CNN distinguishes cores in a given tissue picture, while the second modified CNN takes patches focused at the identified atomic focuses as contribution to anticipate the likelihood that the fix has a place with an instance of PCa repeat. Before applying our modified CNN models, presented the subtleties of the information, which used to build up the proposed PCa repeat model.

### Advantages of Proposed System

The advantages are as follows:

- High accuracy.
- Low time complexity.
- High performance.
- Reduces computational cost.
- Even works with small amount of training data is better than the existing system.
- Deep convolution network [10, 22] based on the concatenation of Exception and ReNet50V2 to improve the accuracy.
- Training technique for dealing with imbalanced image data sets.
- Evaluate our networks on 11,302 chest X-ray images [23].
- Evaluation by using ResNet50V2 and exception on image data set and compared our proposed network with them (Fig. 6.3).

## System Design and Implementation

The basic idea of system design and implementation is to ensure that the COVID-19 patients information built can accommodate for their early prediction [24]. This system design is thus a strategy or forte of portraying the plan, fragments, modules, interfaces and data for a proper structure to satisfy essentials. There are some spread

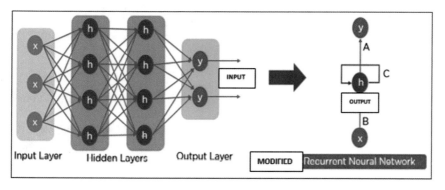

**Fig. 6.3**  Architecture of modified CNN [MCNN]

and joint effort with the data sets in terms of their structure assessment, systems plan and systems building. Execution or proficiency is estimated based on their yield projected by the application. Prerequisite particulars have found to have a significant influence in the investigation of their framework. It is also expected to rest on a great extent with the existing clients of the current framework, through the necessity particulars [24].

## *Implementation and Experimentation Details*

As described above, in the implementation and experimentation of our work, we combined CNN and CNN to achieve better results in terms performance, accuracy and time complexity. The corresponding implementation aspects of systems modules and ECNN algorithms [24, 25] can further be described as following.

## *System Modules*

We call the execution flow of experimentation as the systems module in our work that can be discussed in three modules as following:

- Collecting image data sources.
- Preprocessing image data sets.
- Feature extraction learning.

**Collecting Image Data Sources**: We have taken data sets in UCI, Kaggle and Google websites. For our experimentation [13, 14], we found to have used two types of data sets, to assess the performance risk.

**Structured Data**: Structure data is simply a tabular data that refers to information with a high level of association rows and columns with other data. To that extent, it incorporates within a consistent social database and promptly helpful to search by basic, direct web crawler calculations or other hunt activities. Most of the times, this is commonly visible in .CSV [18, 19] format.

**Preprocessing Image Datasets**

- FEATURES, i.e., attributes through which patients are affected are extracted from the image data sets.
- Preprocessing maybe eliminating duplicate values and adding missing values.
- Each feature importance in affecting the patient can be found using correlation analysis or in max pooling stages.
- In case of unstructured, data needs to be processed to structure data with target class.

## ECNN Algorithm

In view of achieving more accuracy, performance and time complexity, we are forced to extend CNN, to an extended CNN (ECNN). The ECNN [19, 22] is a type of NN where the output from the previous step is fed as input to the current step. CNNs are mainly used for sequence classification-sentiment classification [10, 11, 26] and video classification. To derive the above advantages, we made some changes to CNN and obtain ECNN through a sequence [20, 21] of steps as:

Step 1: Decides how much past data it should remember.
Step 2: Decides how much this unit has to be added to the current state.
Step 3: Decides what part of the current cell state makes it to the output.

Based on the COVID-19 data set, i.e., a total 960 images, our experimentation comprised of the following thirteen steps:

Step 1: Import the required libraries.
Step 2: Import the training dataset.
Step 3: Perform feature scaling to transform the data.
Step 4: Create a data structure with 60-time steps and 1 output.
Step 5: Import Keras library and its packages.
Step 6: Initialize the ECNN.
Step 7: Add the LSTM layers and some dropout regularization.
Step 8: Add the output layer.
Step 9: Compile the ECNN.
Step 10: Fit the ECNN to the training set.
Step 11: Load the COVID-19 test image data for 2020.
Step 12: Get the predicted COVID-19 for 2020.
Step 13: Visualize the results of predicted and real COVID-19.

Thus, the perforce of our algorithm is found to be of more accuracy, consuming less execution time, detailing the COVID-19 cases in their early prediction [24, 27].

## Results

See Figs. 6.4 and 6.5.

## *Evaluation Methods*

In order to exhibit and assess the impact of our proposed method on MCNN, we have adopted few strategies as following. Initially, OP, BP, FN and GN are defined on the individual basis, for confusion matrix, to be examined first. At the same time, the quantity of examples is erroneously anticipated as required due to BP. The quantity of cases is accurately anticipated as not required due to GN and finally the quantity of occasions inaccurately anticipated as not required due to FN. At that point, we can get four estimations such as: performance, accuracy, time complexity review and F1-measure which are calculated based on the following formulae:

$$Accuracy = \frac{OP + ON}{OP + BP + ON + BN}$$

$$Precision = \frac{OP}{OP + BP}$$

$$Recall = \frac{OP}{OP + BN}$$

```
tbarua1@ubuntu:~$ python3 rnn_image.py
/home/tbarua1/.local/lib/python3.6/site-packages/tensorflow/python/framework/dtypes.py:516: FutureWarning: Passing (type, 1) or '1type' as a synonym
of type is deprecated; in a future version of numpy, it will be understood as (type, (1,)) / '(1,)type'.
  _np_qint8 = np.dtype([("qint8", np.int8, 1)])
Step 1, Minibatch Loss= 2.6693, Training Accuracy= 0.148
Step 200, Minibatch Loss= 2.1817, Training Accuracy= 0.281
Step 400, Minibatch Loss= 1.9373, Training Accuracy= 0.453
Step 600, Minibatch Loss= 1.8595, Training Accuracy= 0.414
Step 800, Minibatch Loss= 1.6794, Training Accuracy= 0.500
Step 1000, Minibatch Loss= 1.5005, Training Accuracy= 0.539
------------------------------------------------------------------------
Step 7000, Minibatch Loss= 0.5716, Training Accuracy= 0.836
Step 7200, Minibatch Loss= 0.6380, Training Accuracy= 0.805
Step 7400, Minibatch Loss= 0.6344, Training Accuracy= 0.812
Step 7600, Minibatch Loss= 0.5249, Training Accuracy= 0.828
Step 7800, Minibatch Loss= 0.5086, Training Accuracy= 0.836
Step 8000, Minibatch Loss= 0.4893, Training Accuracy= 0.828
Step 8200, Minibatch Loss= 0.5642, Training Accuracy= 0.805
Step 8400, Minibatch Loss= 0.4294, Training Accuracy= 0.852
Step 8600, Minibatch Loss= 0.5157, Training Accuracy= 0.852
Step 8800, Minibatch Loss= 0.4112, Training Accuracy= 0.867
Step 9000, Minibatch Loss= 0.4896, Training Accuracy= 0.859
Step 9200, Minibatch Loss= 0.4786, Training Accuracy= 0.859
Step 9400, Minibatch Loss= 0.4639, Training Accuracy= 0.859
Step 9600, Minibatch Loss= 0.4180, Training Accuracy= 0.852
Step 9800, Minibatch Loss= 0.4679, Training Accuracy= 0.891
Step 10000, Minibatch Loss= 0.4221, Training Accuracy= 0.875
Optimization Finished!
Testing Accuracy: 0.8671875Step 7000, Minibatch Loss= 0.5716, Training Accuracy= 0.836
```

**Fig. 6.4** Number epochs with accuracy and loss

**Fig. 6.5** Execution flow of COVID-19 dataset

$$\text{F - measure} = \frac{2 \times \text{Precision} \times \text{Recall}}{\text{Precision} + \text{Recall}}$$

Considering every data set and all parameters into consideration, for demonstrating chances of illness, the precision of hazard expectation is expected to rely upon the assorted variety highlight of the medical clinic information. That is, the improvement in the element portrayal of the ailment corresponds to higher the exactness. Our experiment surfaced the precision rate at 90.54% to all, which are under 'more likely' assess the exposure (Fig. 6.6).

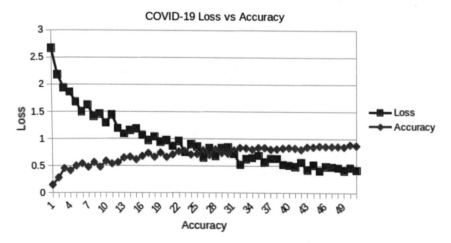

**Fig. 6.6** Data loss versus accuracy

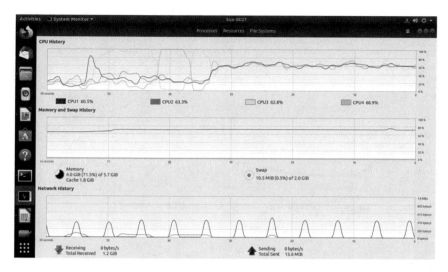

**Fig. 6.7**  Execution time between COVID-19 virus dataset versus number processors

**Time Complexity/Execution Time**: It is realized that our approach consumes 50% less time when compared to other existing techniques. This time can be further minimized using graphic processing unit (GPU) and tensor processing unit (TPU). This entire work execution time is also depended on the system performance. Finally, the system performance in turn depends on the system software, system hardware and available space.

**Input**: As mentioned earlier, our experiment is for considering 960 image data set for the input, from the image database.

The graphical image is thus obtained as shown below (Figs. 6.7 and 6.8):

## Conclusions and Future Possibilities

This research presents a seminal results toward groundbreaking and modern approach for early diagnosis of COVID-19 detection. It primarily concludes a particular mechanism of the proposed COVID-19 early detection system. The concluding analysis of this work is based on evaluating different acoustic features of cough, breath and speech voices. This research elegantly compares and concludes a patients' voice inconvenient accuracy and is found to be proportional to his/her cough and breathe sounds. The research also surfaces the main reason behind these in-efficient preliminary results, as time constraints and computing power. The concluding image database and sound data set is found to be comparatively small and lacks control group data on health database topics, especially in comparing with other patients suffering from other respiratory [4] problems. Eventually, this research data analysis

A. Srinivasulu et al.

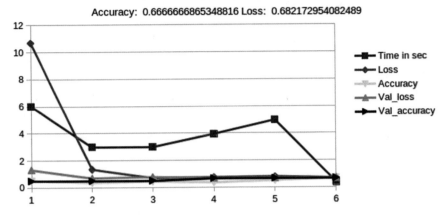

**Fig. 6.8** COVID-19 data size versus accuracy

and observation basis of patient's cough and breathe are found to be more effective factors to diagnose infections. As a result, based on some peoples' quarantine experience, it surfaced that the production of chat bots is soon possible to provide mental support and used as an aid in controlling the anxiety/expected disorders.

This work is based on several typical prediction [25, 28] algorithms and proposes a new framework to create high accuracy and high convergence speed. The proposed modified CNNs algorithm is found to be adoptable for structure and unstructured data from hospitals. The core objective of this proposed system is to enhance the accuracy and performance of predictive model. This work infers that the used COVID-19 data set undauntedly increases the performance and accuracy of most algorithms, as long as MCNN and linear regression lead others.

This work demarks the classified training data set into 8 successive phases. Out of 960 images, we have found, 149 COVID-19 positive, 234 pneumonia positive and 250 normal/negative. In terms of numbers, each class is found to be approximately equal to others, reflecting our proposed MCNN self-learn COVID-19 class characteristics. Especially so, it is not only based on the features but also from other two classes. Among 960 images of our training data sets, the remaining 24 images were allocated for evaluating the network. Since this model is based on a huge number of images, it is made to achieve better accuracy. Thus, we conclude that this model achieved an average accuracy of 89.62% and 91.54% sensitivity for COVID-19 class; with an overall accuracy equal to 88.4% between 2 ±ve folds.

As a part of future work possibilities, this trained MCNN can be made available for public, invariably help better medical diagnosis. Currently with the existing few regular prediction [24] algorithms, our projected algorithms exactness/accuracy is at 91% with an assembly speed. In the event, when the larger data sets are made available from COVID-19 patients, it further pushes this model's accuracy as well as the neural network. In view of fast growing trends in NN advanced techniques, it is certainly possible to achieve better accuracy.

# References

1. Pirouz, B., Shaffiee Haghshenas, S., Pirouz, B., Shaffiee Haghshenas, S., Piro, P.: Development of an assessment method for investigating the impact of climate and urban parameters in confirmed cases of covid-19: a new challenge in sustainable development. Int. J. Environ. Res. Publ. Health **17**(8), 2801 (2020)
2. Univers1. Everything about the COVID-19 virus. https://medicine-and-mental-health.xyz/arc hives/4510; 2020–04–12
3. Pirouz, B., Shaffiee Haghshenas, S., Shaffiee Haghshenas, S., Piro, P.: Investigating a serious challenge in the sustainable development process: analysis of confirmed cases of covid-19 (new type of COVID 19 virus) through a binary classification using artificial intelligence and regression analysis. Sustainability **12**(6), 2427 (2020)
4. Barua, T.: Machine Learning with Python. ISBN: 3110697165, 9783110697162 De Gruyter Stem
5. Narin, A., Kaya, C., Pamuk, Z.: Automatic detection of coronavirus disease (Covid-19) using x-ray images and deep Convolutional neural networks (2020)
6. Li, L., Qin, L., Xu, Z., Yin, Y., Wang, X., Kong, B., Bai, J., Lu, Y., Fang, Z., Song, Q., et al.: Artificial intelligence distinguishes Covid-19 from community acquired pneumonia on chest ct. Radiology, 200905 (2020)
7. McIntosh, K.: COVID-19 virus disease 2019 (COVID-19): epidemicology, virology, clinical features, diagnosis, and prevention. (2020-04-10)
8. Jiang, F., Deng, L., Zhang, L., Cai, Y., Cheung, C.W., Xia, Z.: Review of the clinical characteristics of COVID-19 virus disease 2019 (Covid-19). J. Gen. Intern. Med.,15 (2020)
9. Sun, D., Li, H., Lu, X.X., Xiao, H., Ren, J., Zhang, F.R., Liu, Z.S.: Clinical features of severe pediatric patients with COVID-19 virus disease 2019 in Wuhan: a single center's observational study. World J. Pediatr., 1–9 (2020). (WHO. https://www.who.int; 2020-04-10)
10. Wang, S., Kang, B., Ma, J., Zeng, X., Xiao, M., Guo, J., Cai, M., Yang, J., Li, Y., Meng, X., et al.: A deep learning algorithm using CT images to screen for COVID-19 virus disease (Covid-19) (2020)
11. Wang, L., Wong, A.: Covid-net: a tailored deep convolutional neural network design for detection of Covid-19 cases from chest radiography images (2020)
12. Barua, T., Dr. Doshi, R., Hiran, K.K.: Mobile Applications Development with Python in Kivy Framework. ISBN: 3110689383, 9783110689389 De Gruyter Stem
13. An, P., Chen, H., Jiang, X., Su, J., Xiao, Y., Ding, Y., Ren, H., Ji, M., Chen, Y., Chen, W., et al.: Clinical features of 2019 novel COVID-19 virus pneumonia presented gastrointestinal symptoms but without fever onset. 2020
14. Song, F., Shi, N., Shan, F., Zhang, Z., Shen, J., Lu, H., Ling, Y., Jiang, Y., Shi, Y.: Emerging 2019 novel COVID-19 virus (2019-ncov) pneumonia. Radiology, 200274 (2020)
15. Litjens, G., Kooi, T., Bejnordi, B.E., Setio, A.A.A., Ciompi, F., Ghafoorian, M., Van Der Laak, J.A., Van Ginneken, B., Sanchez, C.I.: A survey on deep learning in medical image analysis. Med. Image Anal. **42**, 60–88 (2017)
16. Wang, L., Wong, A.: COVID-Net: a tailored deep convolutional neural network design for detection of COVID-19 cases from chest x-ray images, pp. 1–12, May 2020
17. Cheng, J.-Z., Ni, D., Chou, Y.-H., Qin, J., Tiu, C.-M., Chang, Y.-C., Huang, C.-S., Shen, D., Chen, C.-M.: Computer-aided diagnosis with deep learning architecture: applications to breast lesions in us images and pulmonary nodules in CT scans. Sci. Rep. **6**(1), 1–13 (2016)
18. Lakshmanaprabu, S., Mohanty, S.N., Shankar, K., Arunkumar, N., Ramirez, G.: Optimal deep learning model for classification of lung cancer on CT images. Future Gener. Comput. Syst. **92**, 374–382 (2019)
19. Zreik, M., Lessmann, N., van Hamersvelt, R.W., Wolterink, J.M., Voskuil, M., Viergever, M.A., Leiner, T., Isgum, I.: Deep learning analysis of the myocardium in COVID-19ry CT angiography for identification of patients with functionally significant COVID-19ry artery stenosis. Med. Image. Anal. **44**, 72–85 (2018)

20. Rahimzadeh, M., Attar, A., et al.: Sperm detection and tracking in phase-contrast microscopy image sequences using deep learning and modied csr-dcf (2020)
21. Yang, R., Li, X., Liu, H., Zhen, Y., Zhang, X., Xiong, Q., Luo, Y., Gao, C., Zeng, W.: Chest CT severity score: an imaging tool for assessing severe Covid-19. Radiol. Cardiothoracic Image. **2**(2), e200047 (2020)
22. Farooq, M., Hafeez, A.: COVID-ResNet: a deep learning framework for screening of COVID19 from radiographs (2020)
23. Afshar, P., Heidarian, S., Naderkhani, F., Oikonomou, A., Plataniotis, K.N., Mohammadi, A.: COVID-CAPS: a capsule network-based framework for identification of COVID-19 cases from x-ray images, pp. 1–5, April 2020
24. Srinivasulu, A., Pushpa, A.: Disease prediction in big data healthcare using extended convolutional neural network techniques. Int. J. Adv. Appl. Sci. (IJAAS) **9**(2), 85–92 (2020). https://doi.org/10.11591/ijaas.v9.i2.pp85-92. ISSN: 2252-8814
25. Srinivasulu, A., Chanakya, G.M.: Health monitoring system using integration of cloud and data mining techniques. HELIX Multidiscip. J. Sci. Exp. **7**(5), 2047–2052 (2017). ISSN 2319-5592 (Online)
26. Gozes, O., Frid-Adar, M., Greenspan, H., Browning, P.D., Zhang, H., Ji, W., Bernheim, A., Siegel, E.: Rapid AI development cycle for the COVID-19 virus (Covid-19) pandemic: initial results for automated detection and patient monitoring using deep learning CT image analysis (2020)
27. Srinivasulu, A., Rajesh, B.: Improving the performance of KNN classification algorithms by using Apache Spark. i-manager's J. Cloud Comput. **4**(2) (2017)
28. Ronne Berger, O., Fischer, P., Brox, T.: U-net: convolutional networks for biomedical image segmentation. In: International Conference on Medical Image Computing and Computer-Assisted Intervention. Springer, pp. 234–241 (2015)

## Chapter 7
# A Cyber Physical System Model for Autonomous Tolling Booths

**Shyamapada Mukherjee**

## Introduction

Time is very precious and people hate to wait for anything in this era of faster life and technology. Mainly people do not like to stand in a queue. Whenever it is a matter of driving and reaching somewhere, waiting becomes most irritating thing. But, most of the time, we are compelled to wait in queues at different tolling offices in highways, whereas we remain in hurry. Due to the human interventions, the process of tax collection at toll offices becomes time consuming and a long queue of vehicle is made up. This increases the waiting time more. Using simple but sophisticated and resilient system can reduce the effort and timing of tax collection at these offices.

Like many cities in the world, India is one of the fastest growing country. The road infrastructure of India is growing very fast. As per the information provided in the government Web site [1], 561 toll plazas are working in India as of now. Almost at all the toll plaza, government collects toll through various contractors. This requires human involvements. With the increasing number of highways and national highways, this number will go upto 1000 within vary short period of time. Therefore, an autonomous tax collection system is needed which work on behalf of government and remove the middleman contractors. This system will make the process simple and will increase the tax collection percentage. The increased collected tax can be used to improve road and the system.

The demand for new ideas with recent technology for automatic toll gate and toll collection has enforced various researchers to give innovative ideas. Many research papers have been published in the direction of autonomous toll gate and toll collection to reduce the unwanted queues of vehicles and unnecessary waiting time at toll plazas. Each paper has its own idea and has some drawbacks also. Many papers have not examined the feasibility of the ideas with respect to the available infrastructure and technology implementation in a country like India. Most of the national papers on this problem have identified the idea but fail to give clear solution in perspective of India.

S. Mukherjee (✉)
National Institute of Technology Silchar, Silchar, Assam 788010, India
e-mail: shyama@cse.nits.ac.in

© The Author(s), under exclusive license to Springer Nature Singapore Pte Ltd. 2022          83
Ch. Satyanarayana et al. (eds.), *Machine Learning and Internet of Things for Societal Issues*, Advanced Technologies and Societal Change,
https://doi.org/10.1007/978-981-16-5090-1_7

A secure toll collection strategy has been presented in [2] when the vehicle on the move. The VANET technology has been used for secure payment using blind coins [3]. A similar kind of idea has been proposed for automatic toll gate system using VANET in [4]. This paper has given an idea of automatic toll collection but not considered many aspects like Internet infrastructure and connection in India. It also missed the behavior of Indian drivers who always fail to behave according to the rules. So missing these important points, a system like autonomous toll plaza concept will work with existing policies in country like India. In [5], an approach for autonomous toll collection is presented. OpenCV tool has been used to identify number plate of a vehicle, and toll is deducted from the owners banks account. The idea in this paper is not complete and has not tested the feasibility in Indian perspective. In [6], a very vague idea of number plate recognition has been talked about, but the idea does not say about toll collection, etc. There are many Indian papers which have been written with very unclear idea for automatic toll collection to achieve the main objectives of the systems. Similar kind of idea has been devised in [7] using RFID for vehicle identification. This paper does not give a complete idea for implementing autonomous toll plaza. A blockchain-based approach using Ethereum has been presented in [8] to implement a transparent toll collection to remove the middle man contractors and malicious handling of toll tax. But this is also a very initial idea and not fully tested its feasibility and importance of using blockchain. The paper in [9] proposed an initial idea of vehicle identification for toll collection. A computer vision technique on OpenCV has been used for identification of vehicles. A geo-fencing approach has been used in [10] based on GPS and GPRS for automatic toll collection from the vehicle owner's account. This idea is not feasible in Indian perspective because of Internet infrastructure and privacy issues. An electronic toll collection system has been proposed in [11] for India. RFID technology has been used to identification of vehicle along with auto-deduction toll fee which is implemented with Arduino Mega. An idea of identifying vehicle at toll plaza using LiFi (light fidelity) technology has been devised in [12]. A very complex hardware infrastructure is required and nothing said about the toll fees collection method. The paper [13] presented an electronic toll collection at Indian toll plazas with the help of RFID technology for vehicle identification. A cloud-based tax collection is also talked about. An toll tax automation and monitoring system has been proposed in [14]. The RFID technology has been used for vehicle identification and tax collection using Android application. In [15], RFID-based identification technique has been discussed where tax collected automatically by central controlling process. In [16], an electronic toll collection approach has been presented using the application ZigBee and RFID. Though it does not need human intervention, but in reality, the existing transport system of India will not be benefited from it. The paper [17] presented a smart toll collection system using vehicle identification from the previously stored database to generate bill. [18] proposed an automated toll collection approach by identifying vehicle using barcode-laser scanning when vehicles are on the move. The paper [19] presented an approach to remove manual toll collection using ZigBee and GPS. The tax is deducted from the account of the owner of the identified vehicle. The paper [20] proposes an idea to implement an automatic toll tax collection system using RFID and GSM.

Many researchers outside of India have published various research articles on electronic toll gate and toll collection systems. In the paper [21], two different tolling strategies have been presented. The two approaches, feedback control and another by learning fashioned, have been delivered by the proof of concept for toll rate determination. The paper [22] represents an image processing-based technique for number plate identification of vehicles at toll booths to collect toll tax in Pakistan. The system calculates the toll tax and generates the bill which is stored in a database. An economical toll tax system is presented in [23] which uses an image processing technique using Raspberry Pi model to reduce the cost of making a system. A review on European truck tolling schemes has been presented in [24]. It also depicts an assessment of the impact of the system on the logistics. An automated toll collection and pricing strategy for European countries has been presented in [25]. A RFID-tagged vehicles are identified by road-side systems using short-range communication which calculated the toll price the same time using the rules set by the authority. In the paper [26], an automatic toll booth system is presented. The vehicles are identified using RFID tag fitted with the vehicles, and the toll tax is automatically deducted from the owner's account. The paper [27] presents a comparative study on three digital toll collection systems in Surabaya. A comparison on service performance has been measured in this paper. A receiver autonomous integrity monitoring (RAIM)-based electronic toll collection systems has been presented in [28]. The RAIM algorithms run on Global Navigation Satellite Systems (GNSS) to identify the vehicles on the run reduce the traffic at toll booths. A gray scale imaging technique has been used in [22] to identify the number plates of vehicles at the tool plaza for automatic toll collection. An integrated system deducts the tax form owner's account. Another paper on automatic toll collection system using RFID tag is presented in [29]. This paper also suggests to block a vehicle by not lifting the toll gate if the tolling authority detects any rule violation by the driver.

Many papers have been published on electronic toll collection system in Indian tolling booth perspective. But most of the of ideas are incomplete and not suitable for existing technological infrastructure in terms of Internet facility, privacy, and automatic tax collection or deduction from account of the owners of the vehicles. Some of the papers and patents have been discussed in the previous paragraphs. Those papers presented automatic toll plazas in foreign countries. Our main objective is to present a feasible solution for autonomous tolling system in India and other countries considering various technological constraints and existing infrastructure.

An intelligent cyber physical system model is introduced in this paper. This system will identify the type of vehicles and recognize the number plates. Based on the identification and load on it, toll tax will be calculated. A toll ticket will be generated for each vehicle and based on the previous information. The toll tax will be deducted from the recharged amount of the vehicle. After that, toll gate will be lifted. The detailed operation of the systems is discussed in the following sections.

The rest part of the paper is organized as in the following: Section "Problem Definition" describes the problem along with various cases arise at toll booths. In

Section "Cyber Physical System for Autonomous Toll Booths (CPSATB)," the proposed system is presented. A proof of concept has been presented in Section "A Proof of Concept for CPSATB." Section "Conclusion" concludes the paper.

## Problem Definition

With improved road and highway infrastructure, the challenges of managing traffics at different toll booths have become herculean task in a country like India. Long queues of vehicle at toll booths consume the valuable time of the people, and along with that it avoids the main aim of constructing many lanes highways. Electronic tolling system is a demand of the situation. Such systems can identify vehicles and deduct toll charges from the previously recharged amount against the vehicle. After deducting the charges, the bar at the toll gate is lifted automatically. This altogether reduces the waiting time at toll booths and removes the human interventions.

The ministry of transport of India has made FASTag and digital payment mandatory for toll at various toll booths via FASTag and digital payments for toll via a radio frequency identification (RFID) chip card sticker pasted on the vehicle's windscreen. The aim of this idea is to reduce the waiting time of vehicles at toll booths. It also ensures the reduced traffic congestion at toll booths which helps smooth flow of traffic. Congestion at toll booths presently results in an estimated annual loss due to wastage of fuel of over Rs 12,000 Crores. FASTag has shown many merits. In addition to reducing the waiting time, it helps in increasing the total collected revenue from toll. With the cash payments systems, toll booth operators could set up under the table deals with regular users and transporters. But with FASTag card reader system, barriers at toll booths are connected and controlled by the card reader system, and the chances of a vehicle driving through without paying toll are negligible. Digital/Electronic payment means a direct deposit in a bank account, which reduces mistakes in counting and collection. In addition to this, cash collection has danger of looting and thefts. To prevent this, most toll plazas, especially in states with law and order issues, have armed guards posted round the clock. An electronic payment system would reduce the need for such private armed guards and display of muscle power.

Though there are many benefits of using FASTag and similar kind of systems for electronic toll system, many challenges are associated with it. There are two sides to every coin. Here, very basic challenges are presented, which many drivers face with the FASTag and RFDI technology- based electronic tolling systems.

1. **P1**: The most common concern that people face is the loss or theft of FASTags pasted on wind screen of vehicles.
2. **P2**: Due to technical errors, toll charges are deducted twice or more from owners of the vehicles.
3. **P3**: Sometimes FASTag may not work that means vehicle going through the RFID scanner, but it is not able to detect the FASTag.

4. **P4**: Vehicles without FASTag may enter in the lane for FASTag vehicles, which increases waiting time significantly. This is one of the major problem that arises in India. It voids the purpose of the electronic toll collection.

   There need to be stringent measures to ensure that the entry way into the designated FASTag lanes also has a scanner. This scanner will be the means to detect the card securely attached on the windscreen of vehicles. This will ensure that entry is allowed only to automobiles that have FASTag. And when these measures are in place, the user satisfaction and time-saving opportunities will grow manifold.

5. **P5**: An event that the users do not maintain sufficient balance in their corresponding FATag account/wallet. Hence, users try to pass through the toll gates without adequate balance in their FASTag wallet are not able to avail FASTag services. The system enforces them to pay the toll in cash, which increases length of the waiting queue. This indirectly causes greater fuel consumption and carbon emission.

These problems need immediate technological feasible solutions to achieve the aims of the electronic toll collection. In this paper, we have taken care of all the real-life problems at various toll plazas and proposed technical feasible solutions.

## Cyber Physical System for Autonomous Toll Booths (CPSATB)

In this section, an intelligent cyber physical system for electronic toll collection has been presented where the system is built up on the real issues raise at toll booths and their operations. Figure 7.1 describes the block diagram of the proposed CPS system. This system works autonomously without any human intervention. The working principle of the model presents that how this system considers all common real problems raised at a toll booths even they are fully electronic systems. The challenges are identified and described in the previous sections. The operations of each of the components of the system are described in next part of this section.

The system is conceptualized as a three layers systems: (1) input layer, (2) control layer, and (3) output layer. In a cyber physical system, sensors are the components which interact with the physical world in the input layer and provides data to the control unit or processing unit. The control unit send the information to the actuators to react accordingly in the output layer.

### *Input Layer of CPSATB*

The input layer of the proposed CPS system is consisting of three input components: (i) a RFID unit, (ii) a deep learning model for computer vision, (iii) an weighing component.

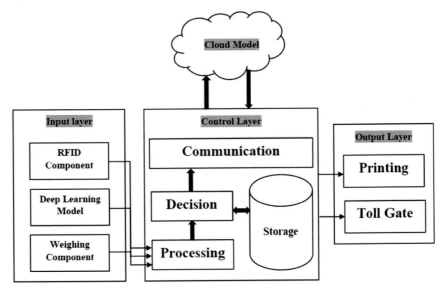

**Fig. 7.1** Block diagram of a cyber physical system for autonomous toll booths (CPSATB)

**RFID Component**: The RFID-integrated component work in similar fashion like FASTag. A radio frequency chip will be attached with each vehicle. When a vehicle tagged with RFID passes through a lane of a toll booth, a stationary RFID scanner system will detect the information from the RFID tag. The scanner then passes the information to the processing unit of the next layer of the CPS. The information will consist of an identity code of the vehicle.

**Deep Learning Model**: This is an additional part of the system to identify a vehicle by reading the numbers on the number plate of a vehicle. A high-resolution camera is fixed somewhere on a fixed bar near each lane of a toll both. This camera takes five photographs of the number plate and process them immediately using a deployed deep learning model to retrieve the number of the vehicle. In addition to that, a second camera is installed on the same bar to take a photograph of the vehicle. From this photograph, the vehicle type is identified using the deployed deep learning model. After that, those information are forwarded to the processing unit of the control layer. Identification of type of vehicles is needed in the cases where the toll charge is decided based on the type of the vehicles.

This component helps reducing the problem of non-identification of vehicle due to non-working of RFID tags or lost or theft tag. The second method of vehicle identification by deep learning model removes the problem identified as **P1** and **P3** in the problem definition section.

**Weighing Component**: This component measures the weight of vehicles. The corresponding sensors placed under the ground where the vehicle stop for toll at a toll booth. This measurement is used to decide the toll tax. The proposed CPS system can be customized to use the vehicle type identification and weighing component to decide tax or over weight charge. The information from the weighing component is passed to the processing unit of the control layer.

## Control Layer of CPSATB

The control layer of the proposed CPS system is consisting of four components: (i) a processing unit, (ii) a decision unit, (iii) an communication unit, and (iv) a storage unit. Each component has a vital role in the CPS for autonomous toll booth.

**Processing Unit**: The processing unit gets the input data from the different components of the input layer. It then processes the information for further use. Based on the customization done on the system, information are passed to the decision component of the control layer for the decision to be made.

**Decision Unit**: The processed information come to this unit for the decision of toll amount based on the vehicle type and wight or both. In this customizable system, authority can set any parameters for deciding toll charges. The decision unit connects the communication unit to deduct the toll tax from the account of the owner of the vehicle. Here, few situations can arise due to Internet connectivity and vehicle recharge account details stored in the cloud.

1. **Case 1**: *Internet connection is not available or poor connectivity to complete transaction*

   **CPS Solution**: If the Internet connectivity is not available or the connection speed is poor to complete the transaction, then the decision unit stores the details of the vehicle along with the toll to be collected in the storage unit and sends instructions to the printing unit for printing the toll slip mentioning that due to technical issue the toll will be deduced after sometime or similar kind of message as required by authority. At the same time, the decision unit sends an instruction to the toll gate system for lifting the gate so that vehicle can go through. At later time, when the Internet connection is up, the all due tax stored in the storage unit is deducted from the corresponding accounts by the decision unit with the help of communication unit and cloud unit.

2. **Case 2**: *Good Internet connection*

   **CPS Solution**: If the there is no Internet connectivity issue, through communication unit, the decision unit communicates with the cloud system where data for all the registered vehicle along with the all other information regarding recharge account balance etc. are stored. The cloud system deducts the toll amount from the corresponding account and sends a transaction completion message to the decision unit. The decision unit sends an instruction to the printing unit to print a toll slip. At the same time, the decision unit sends an instruction to the Toll gate system for lifting the gate so that vehicle can go through.

3. **Case 3**: *Good Internet connection but account balance is low*

   **CPS Solution**: If the corresponding account balance for a vehicle is found low, i.e., the amount in the account is less than the toll charge or zero, then the cloud stores these information and sends a message of non-payment of toll to the decision unit. The decision unit sends an instruction to the printing unit to print a toll slip mentioning the non-payment of toll. At the same time, the decision unit sends an instruction to the toll gate system for lifting the gate, so that vehicle can go through.

   *Policy for the non-payment of toll due to low account balance*:

The non-payment toll and its related information about the vehicle, date, time, toll booth id, etc., will be stored in the cloud. But the corresponding vehicle will not be stopped to avoid queuing hazard at toll booth. This will remove the long queue problem in toll booths and saves time and fuel wastage which is a common problem in real scenario. There would be two options for the vehicle owners to pay their unpaid toll. A vehicle owner can pay the due toll charge with a certain penalty on the specified payment portal provided by the authority. To make the system more strict, a policy should be adopted. Under this policy if any vehicle avoids payment of due toll, the toll will be charged by enforcing the vehicle owner to have a toll clearance certificate at the time of renewing insurance of the vehicle. This certificate can be obtained only by paying all the due toll tax or have no due. Adoption of this policy will help the toll system robust than existing one.

**Communication Unit**: This unit is the part of the second layer of the CPS. This will make communication with the cloud and maintain different communication protocol specifically for this system.

**Storage Unit**: The storage unit in the CPS system will store the temporary information which are generated due to the poor or unavailable communication with the cloud. This information will be transferred to the cloud when the communication link is up. This information in the cloud is stored as due toll for the corresponding vehicle. But there should be policy, so that there will be no extra penalty for this type of due toll incidents.

## Cloud Model

This is the central information storage and processing unit for the various toll booths in a country. It has direct communication with the toll booths and the CPS systems installed at a different lanes of toll booths. The cloud store the details of all the registered vehicles along with the recharge account balance for payment of toll charges. This will also stores the due toll of different vehicles. This information can be accessed for any registered vehicle through a dedicated Web portal to pay due tolls and to generated toll clearance certificate required for renew insurance of vehicles.

To solve the problem P2, the processing unit of the cloud will check whether any vehicle has been charged twice or more times in a very short period of time (in minutes). From the timestamp values of the transactions, it can be easily identified. The extra deducted amount will be refunded back to the corresponding vehicle accounts. This will remove the problem of double deduction.

## *Output Layer of CPSATB*

The output layer of the proposed CPS system is consisting of two components: (i) a printing unit, and (ii) a toll gate unit.

**Printing Unit**: The printing unit is a part of the third layer of the proposed CPS. It is responsible of printing the toll slip. The decision unit of the control layer sends the details of toll deducted or not deducted along with the required message to be printed. This prints the toll slip and then sends an instruction to the decision unit about the completion of the printing job.

**Toll Gate Unit**: This is the second component of the third layer which is responsible for lifting and closing the toll gate based on the received instruction from the decision unit. The decision unit sends an instruction to the toll gate unit to lift the gate after receiving a printing completion message from the printing unit. After a vehicle passes the toll gate, the decision unit receives a message from the weighing unit regarding the change in load (weight) from certain value to nearly zero value. The decision unit passes an instruction to the toll gate unit for closing its bar.

## A Proof of Concept for CPSATB

The proposed cyber physical system model has been implemented using C++ programming language to present its feasibility and efficiency through a proof of concept. This takes the proposed model from TRL (Technology Readiness Level) level 2 to TRL level 3 which has been described in this section.

Each unit (Fig. 7.1) in the proposed model has been conceptualize as a method or service called by some other methods. We have tried to present a comparative performance analysis with the existing electronic toll collection system and associated policies. Figure 7.2 shows the work flow of the various methods representing the proposed CPS model. The model has been tested with all possible input or cases. The corresponding outputs are presented in terms of action taken.

In Table 7.1, five possible cases have been reported. These cases or issues are very real which arise in toll booths. The cases reported in the table represent the following situations:

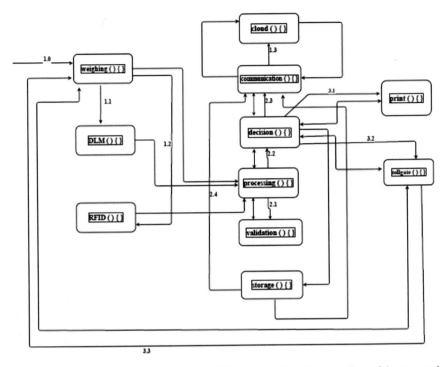

**Fig. 7.2** Interaction among various methods which are mimicking the operations of the proposed CPS for autonomous toll booths

**Table 7.1** Comparative analysis of the proposed CPS for autonomous toll booths with respect to the existing electronic toll collection systems

| Cases | Existing ETC | | | | Proposed CPS | | | |
|---|---|---|---|---|---|---|---|---|
| | Proc. T. | WT | WoF | MPI | Proc. T. | WT | WoF | MPI |
| Case 1 | 1 | 1 | 1 | 0 | 1 | 1 | 1 | 0 |
| Case 2 | 5 | 5 | 5 | 5 | 1 | 1 | 1 | 0 |
| Case 3 | 5 | 5 | 5 | 5 | 1 | 1 | 1 | 0 |
| Case 4 | 5 | 5 | 5 | 5 | 1 | 1 | 1 | 0 |
| Case 5 | 5 | 5 | 5 | 5 | 1 | 1 | 1 | 0 |

1. Case 1: Everything is normal or no abnormality identified.
2. Case 2: RFID tag is missing or torn, i.e., vehicle is registered in the corresponding system and has a balance in the account.
3. Case 3: Internet connection is poor and transaction not completed due to that.
4. Case 4: Account balance of the vehicle is low.
5. Case 5: Vehicle is not registered in the system and no account maintained.

We have presented a proof concept for the proposed systems. The system is not designed physically. Hence, actual comparison of the systems with electronics toll collection systems in practice is not possible. But, the cases identified help to measure the performance of two systems with the dummy values. To compare these two systems, we have used a range of values form 0 to 5 where 0 is the best and 5 are worst for the parameters considered here. The table columns "Proc. T.," "WT," "WoF," and "MPI" represent the processing time, waiting time of a vehicle, wastage of fuel, and man power involvement. It can be easily understand that for all the cases, the proposed CPS will have no human intervention. The proposed system will save money, fuel, and time altogether.

Figure 7.2 depicts the working principle of the proposed CPS. The interaction among various methods which are mimicking the different components of the CPS is shown.

A proof of concept of the proposed cyber physical system for autonomous toll booths is shown in Fig. 7.3. The five cases identified have been considered and shown how the system works under these situation smoothly. Where the traditional electronic toll collection systems fail to handle these and cause wastage of time and fuel at the same time, human intervention is needed.

## Conclusion

The proposed cyber physical system is a complete solution for implementing an autonomous toll booth system. It helps in removing all set of problems identified in the electronic toll collection system in practice. This system not only abolishes the human intervention completely, it resolves all kind of hazards arise at different toll booth in India or similar type of countries. This system will be able to save the time and fuel which are most precious nowadays. At the same time, it is able to collect each penny from the vehicles passing through the toll gates. This will increase the revenue collected from toll booths.

In this paper, the proof concept presented to show the working principle of the system. The proof of concept proves that this idea is really implementable and saves time and money for the public as well as government. All sort of possible cases have been considered to show that in country like India, it is really important and feasible with the existing technological infrastructure.

In future, will be developing a complete model of the proposed CPS to achieve TRL 5. The proposed system can be developed with more recent advanced technology to achieve zero waiting time at toll gates. We will be working toward this direction in near future.

```
Load is Calculated
Calling RFID Unit
RFID Unit is scanning RFID TAG
Scanning Complete
RFID 6753562912

Calling  Deep Learning model
Deep Learning Unit is Active
Number Plate is being scanned
Vehicle is being scanned
Number Plate is recognized
Vehicle type is identified : small_car
number plate AR36WK2618
Data is being processed
RFID value is being validated
RFID is valid

Vehicle number if being validated
Vehicle number is valid

Toll value is being calculated
TOLL is 8769

TOLL is being deducted
Connecting to retrieve vehicle details
Transaction is complete: Toll tax deducted
Connecting printing unit to print the TOLL slip
Printing toll slip

TOLL slip is printed
Connecting Toll gate to lift the bar
Toll gate is lifting the bar
Toll Gate is down
```

[Case 1]

```
$main
A vehicle has arrived
Weighing Unit is Active
Load is Calculated
Calling RFID Unit
RFID Unit is scanning RFID TAG
Scanning Complete
Calling  Deep Learning model
Deep Learning Unit is Active
Number Plate is being scanned
Vehicle is being scanned
Number Plate is recognized
Vehicle type is identified : small_car
number plate AR36WK2618
Data is being processed
RFID value is being validated
RFID Tag is missing

Vehicle number if being validated
Vehicle number is valid

Toll value is being calculated
TOLL is 8769

TOLL is being deducted
Connecting to retrieve vehicle details
Transaction is complete: Toll tax deducted
Connecting printing unit to print the TOLL slip
TOLL slip is printed
Connecting Toll gate to lift the bar
Toll gate is lifting the bar
Toll Gate is down
```

[Case 2]

```
Load is Calculated
Calling RFID Unit
RFID Unit is scanning RFID TAG
Scanning Complete
RFID 6753562912

Calling  Deep Learning model
Deep Learning Unit is Active
Number Plate is being scanned
Vehicle is being scanned
Number Plate is recognized
Vehicle type is identified : small_car
number plate AR36WK2618
Data is being processed
RFID value is being validated
RFID is valid

Vehicle number if being validated
Vehicle number is valid

Toll value is being calculated
TOLL is 8769

TOLL is being deducted
poor internet connectivity
information is stored temporarily
Toll details are stored for future action

Connecting printing unit to print the toll bil
TOLL slip is printed
Connecting Toll gate to lift the bar
Toll gate is lifting the bar
Toll Gate is down
```

[Case 3]

```
A vehicle has arrived
Weighing Unit is Active
Load is Calculated
Calling RFID Unit
RFID Unit is scanning RFID TAG
Scanning Complete
RFID 6753562912

Calling  Deep Learning model
Deep Learning Unit is Active
Number Plate is being scanned
Vehicle is being scanned
Number Plate is recognized
Vehicle type is identified : small_car
number plate AR36WK2618
Data is being processed
RFID value is being validated
RFID is valid

Vehicle number if being validated
Vehicle number is valid

Toll value is being calculated
TOLL is 8769

TOLL is being deducted
no deduction due to less account balance
Due TOLL is stored for later payment
Connecting printing unit to print the TOLL slip
TOLL slip is printed
Connecting Toll gate to lift the bar
Toll gate is lifting the bar
Toll Gate is down
```

[Case 4]

```
Weighing Unit is Active
Load is Calculated
Calling RFID Unit
RFID Unit is scanning RFID TAG
Scanning Complete
RFID 6753562912

Calling  Deep Learning model
Deep Learning Unit is Active
Number Plate is being scanned
Vehicle is being scanned
Number Plate is recognized
Vehicle type is identified : small_car
number plate AR36WK2618
Data is being processed
RFID value is being validated
RFID is valid

Vehicle number if being validated
Vehicle number is valid

Toll value is being calculated
TOLL is 8769

TOLL is being deducted
Connecting to retrieve vehicle details
no account is found
Due TOLL is stored for later payment
Connecting printing unit to print the TOLL slip
TOLL slip is printed
Connecting Toll gate to lift the bar
Toll gate is lifting the bar
Toll Gate is down
```

[Case 5]

**Fig. 7.3** Proof of concept of the proposed cyber physical system for autonomous toll booths. (a) Case 1: Normal condition. (b) Case 2: RFID-related issues. (c) Case 3: Internet connection issues. (d) Case 4: Account balance issue. (e) Case 5: Vehicle registration issue

# References

1. http://tis.nhai.gov.in/tollplazasataglance?language=en
2. Chaurasia, B.K., Verma, S.: Secure pay while on move toll collection using VANET. Computer Standards Interfaces **36**(2), 403–411 (2014)
3. Popescu, C.: A fair off-line electronic cash system based on elliptic curve discrete logarithm problem. Studies Inform. Control **14**(4), 409–416 (2005)
4. Senapati, B.R., Khilar, P.M., Sabat, N.K.: An automated toll gate system using VANET. In: 2019 IEEE 1st International Conference on Energy, Systems and Information Processing (ICESIP), Chennai, India, pp. 1–5 (2019)
5. Sathya, G., Sangeetha, K., Siddharthan, S., Punj, S.: Smart way toll collection system 1 Multidiscip. Res. 9(3) (2019)
6. Shahare, N., Parsewar, S., Bhange, P., Kshirsagar, D., Sawwalakhe, R.: Toll tax collecting system using optical character recognition. **13**(3s) (2020)
7. Sanghvi, K., Joglekar, A.: Automating the payment of toll tax at toll plazas. Int. J. Computer Sci. Inform. Technol. **6**(3), 2884–2887 (2015)
8. Manjaramkar, N., Naikare, S., Patil, S., Shinde, V., Pawar, A.: Tollblocks—an etherum based toll processing system using blockchain. J. Crit. Rev. **7**(19) (2020)
9. Suryatali, A., Dharmadhikari, V.B.: Computer vision based vehicle detection for toll collection system using embedded Linux. In: 2015 International Conference on Circuits, Power and Computing Technologies [ICCPCT-2015], Nagercoil, pp. 1–7 (2015)
10. Nagothu, S.K. : Automated toll collection system using GPS and GPRS. In: 2016 International Conference on Communication and Signal Processing (ICCSP), Melmaruvathur, pp. 0651–0653 (2016)
11. Chattoraj, S., Bhowmik, S., Vishwakarma, K., Roy, P.: Design and implementation of low cost electronic toll collection system in India. In: 2017 Second International Conference on Electrical, Computer and Communication Technologies (ICECCT), Coimbatore, pp. 1–4 (2017)
12. Singh, D., Sood, A., Thakur, G., Arora, N., Kumar, A.: Design and implementation of wireless communication system for toll collection using LIFI. In: 2017 4th International Conference on Signal Processing, Computing and Control (ISPCC), Solan, pp. 510–515 (2017)
13. Khan, E., Garg, D., Tiwari, R., Upadhyay, S.: Automated toll tax collection system using cloud database. In: 3rd International Conference On Internet of Things: Smart Innovation and Usages (IoT-SIU). Bhimtal **2018**, 1–5 (2018)
14. Christopher, K.K., Arul, X.V.M., Karthikeyen, P.: Smart toll tax automation and monitoring system using android application. In: 2019 IEEE International Conference on Intelligent Techniques in Control, Optimization and Signal Processing (INCOS), Tamil Nadu, India, pp. 1–6 (2019)
15. Singh, S., Garg, P.: IOT-based auto-payment of toll tax. Int. J. Computer Appl. **179**(42), 49–53 (2018)
16. Dhilipkumar, S., Arunachalaperumal, C.: Smart toll collection system using ZIGBEE and RFID. Mater. Today Proc. **24**(3), 2054–2061 (2020)
17. Bohra, V., Prasad, D., Nidhi, N., Tiwari, A., Nath, V.: Design strategy for smart toll gate billing system. In: Nath, V., Mandal, J. (eds.) Proceeding of the Second International Conference on Microelectronics, Computing & Communication Systems (MCCS 2017). Lecture Notes in Electrical Engineering, vol. 476. Springer, Singapore (2019)
18. Raj, U., Nidhi, N., Nath, V.: Automated toll plaza using barcode-laser scanning technology. In: Nath, V., Mandal, J. (eds.) Nanoelectronics, Circuits and Communication Systems. Lecture Notes in Electrical Engineering, vol. 511. Springer, Singapore (2019)
19. Kumar, C., Nath, V.: Design of smart embedded system for auto toll billing system using IoTs. In: Nath V., Mandal J. (eds.) Nanoelectronics, Circuits and Communication Systems. Lecture Notes in Electrical Engineering, vol. 511. Springer, Singapore (2019)
20. Mahalakshmi, P., Puntambekar, V.P., Jain, A., Singhania, R.: Automatic toll tax collection using GSM. In: Shetty, N., Patnaik, L., Nagaraj, H., Hamsavath, P., Nalini, N. (eds.) Emerging

Research in Computing, Information, Communication and Applications. Advances in Intelligent Systems and Computing, vol. 906. Springer, Singapore (2019)

21. Yin, Y., Lou, .: Dynamic tolling strategies for managed lanes. J. Transport. Eng. **135**(2), 45–52 (2009)

22. Soomro, S.R., Javed, M.A., Memon, F.A.: Vehicle number recognition system for automatic toll tax collection. In: International Conference of Robotics and Artificial Intelligence, Rawalpindi, pp. 125–129 (2012)

23. Rathan, K.J., Rahaman, M.S., Sarkar, M.K., Sekh Mahfuz, Md.: Raspberry Pi image processing based economical automated toll system. Global J. Res. Eng. (2013)

24. McKinnon, A.C.: A review of European truck tolling schemes and assessment of their possible impact on logistics systems. Int. J. Logistics Res. Appl. **9**(3) (2006)

25. Blythe, P.: RFID for road tolling, road-use pricing and vehicle access control. In: IEE Colloquium on RFID Technology (Ref. No. 1999/123), London, UK, pp. 8/1-816 (1999)

26. Al-Ghawi, S.S., Hussain, S.A., Al Rahbi, M.A., Hussain, S.Z.: Automatic toll e-ticketing system for transportation systems. In: 2016 3rd MEC International Conference on Big Data and Smart City (ICBDSC), Muscat, pp. 1–5 (2016)

27. Presents a comparative study on three digital toll collection systems in Surabaya. A comparison on service performance has been measured in this paper

28. Salós, D., Martineau, A., Macabiau, C., Bonhoure, B., Kubrak, D.: Receiver autonomous integrity monitoring of GNSS signals for electronic toll collection. IEEE Trans. Intell. Transp. Syst. **15**(1), 94–103 (2014)

29. R. Hossain, M. Ahmed, M. M. Alfasani and H. U. Zaman, "An advanced security system integrated with RFID based automated toll collection system," 2017 Third Asian Conference on Defence Technology (ACDT), Phuket, 2017, pp. 59-64

# Chapter 8
# Sentiment Extraction from English-Telugu Code Mixed Tweets Using Lexicon Based and Machine Learning Approaches

**Arun Kodirekka and Ayyagari Srinagesh**

## Introduction

India is a multi-lingual country [1]. People know a minimum of two languages. In social networks, most of the people are using bilingual text to share their opinions and response. Code mixed text is tremendously increased in social networks because of the usage of bilingual text. Writing one language into another language is known as code mixed. In India, code mixed text is near with the combination of the English language [2]. English-Hindi (EN-HI) Hinglish, English-Bengali (EN-BN) Benglish, English-Telugu (EN-TE) Tenglish, are all Indian official languages with English. Writing of any linguistic units, words, and phrases into English is very easy for the public. As the social networks provide linguistic facilities, code mixed text is increasing.

A. Kodirekka
Department of Computer Science and Engineering, Acharya Nagarjuna University, Guntur, Andhra Pradesh, India

A. Srinagesh (✉)
Department of Computer Science and Engineering, RVR & JCCE, Guntur, Andhra Pradesh, India

Ch. Satyanarayana et al. (eds.), *Machine Learning and Internet of Things for Societal Issues*, Advanced Technologies and Societal Change,
https://doi.org/10.1007/978-981-16-5090-1_8

Sentiment analysis is used to extract the sentiments like positive, negative and neutral from different reviews. Extraction of sentiments from English tweets is the traditional method; sufficient research work is done. A sentiment from the code mixed tweet is a research challenge. Particularly, more researchers focus on localized English-Telugu code mixed languages for analysis. Sentiment extraction from the English-Telugu code mixed tweets contains four steps. In the first step, collect the code mixed text from the social networks, secondly language identification, thirdly transliterate the roman script into Telugu script, and finally, sentiment extraction using NLP lexicon-based approach such as TeluguSentiWordNet [3, 4], and machine learning approach. This work is organized in the following sequence. 'Related Work', 'Methodology', 'Experiments and Results', and 'Conclusion and Future Work' in this chapter.

## Related Work

Sentiment analysis is applicable to predict the stance detection, review analysis, and recommendation system. India is the most populated and multi-linguistic country in the world. In this work, Hindi-Bengali, Bengali-Hindi code mixed data was collected from the social networks. Dataset was constructed for Hindi-Bengali, Bengali-Hindi code mixed data. Mahata et al. [5] analysed the rules of code-switching in the code mixed data such as the English-Hindi corpus. The workflow was extracted from code mixed data, and transliterated into its original form, and annotated with the word pairs of parts-of-speech and recognition of the rules of code switched data [1].

Sentiment analysis for Indian languages (SAIL) for the code mixed tool is constructed to extract the sentiments from the Indian language pairs like Hindi-Bengali, Bengali-Hindi-English. In this chapter, two approaches were given; one is a voting classifier that performs a machine learning model by using linear SVM, logistic regression, and random forest. The second one is the linear SVM method, which is the best method [6].

Language identification at the word level is very important in code mixed text [7]. In this work, a code mixed index was defined.

Code mixed text is mixing of more than one language unit [8]. The bilingualism text is very natural in social networks like India. This work addresses the incomplete spellings and equality words and its solutions that were implemented on the Hindi-English dataset.

Sentiment analysis is a natural language process. Indian languages are the low-resourced languages, and Telugu is one among them. This work improved the accuracy of Telugu sentiment. It contains two-step approaches; objective and subjective. The objective-type sentences have neutral sentiment and do not affect the analysis. The subjective sentence is classified as positive or negative with the support of the TeluguSentiWordNet [3].

In this analysis, they consider Telugu news as the dataset from the different e-news Telugu papers. In the subjective classification, accuracy was improved up to 81% [5].

In Telugu language sentiment analysis, there is no benchmark dataset. Telugu is the regional language that has a low recourse dataset. ACTSA corpus dataset is developed to improve the Telugu sentiment analysis. They also implemented Sentiphrasenet and validated with ACTSA, instead of the SentiWordNet approach [9].

Dr. Sentiment defined psychoSentiWordNet. Human psychology is related to many things like community psychology, civilization, pragmatics, and many more continual intelligent aspects of culture that provide online sentiment extraction [10].

Dr. Sentiment was the online interactive gaming approach. It interacts players with sequence queries; finally, it measures the psychology of the user as well as it is built the SentiWordNet. Dr. Sentiment is used to develop the SentiWordNet for the Hindi, Bengali, and Telugu [11].

## Methodology

The code mixed corpus data was collected from Twitter API, and to extract sentiments from the code mixed data especially Telugu code mixed data, Fig. 8.1 explains the proposed methodology and the following steps.

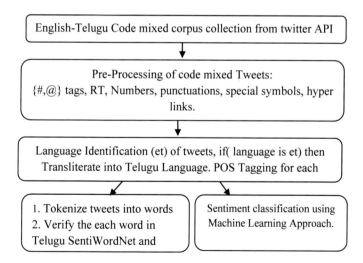

**Fig. 8.1** Sentiment extraction from Telugu-English code mixed corpus using Lexicon and machine learning methods

## Algorithm1: Sentiment Extraction from Telugu Code Mixed Tweets (STCMT)

**Algorithm:** Sentiment classification from the Telugu Code mixed Tweets

**Input:** Telugu-English Code mixed Tweets $\Gamma_{cm}$, Sentiwordnet positive lexicon ($S_p$), Sentiwordnet Negative lexicon ($S_n$), Sentiwordnet positive lexicon ($S_{nl}$)

**Output:** Code mixed Tweet Sentiment (CMTS), Sentiment Score SS

**Notation:** Tweet T, Token_List $T_L$, Token $T_K$, Transliterate_into_Telugu TLT, Sentiment_Positve_score Ps, Sentiment_negative_score $N_s$, Language Identification ($D_L$),

1.  While T in $\Gamma_{cm}$
2.          $T_L \leftarrow$ get_tokens(T)
3.          While $T_k$ in $T_L$
4.                  $D_L \leftarrow$ Identify_Language($T_k$)
5.                  if ($D_L$ == 'et') then
6.                  Telugu_Word=Transliterate_into_Telugu($T_L$)
7.                  If Telugu_Word in $S_p$
8.                      Pos$\leftarrow$pos+1
9.                  else if Telugu_Word in $S_n$
10.                     Neg$\leftarrow$neg+1
11.             Else
12.                     nt$\leftarrow$nt+1
13. //Sentiment score
14.         If pos > neg then
15.                 Sp= pos / (pos+neg)
16. else if pos < neg then
17.                 Sn= neg / (pos+neg)
18.         else
19.                 SS=0
20.         SS=Sp-Sn
21.         End

1.  Corpus collection and annotation.
2.  Pre-processing of the collected corpus of Twitter data.

3.  Language identification and transliteration.
4.  Extracting the sentiments.
1.  **Corpus Collection and Annotation**

Twitter API provides a multi-lingual environment to the users to share the information and give the response. The length of the tweet is only restricted up to 280 characters recently, and there are no other limitations. It allows free form of text in any language, misspellings, incomplete spellings, equality words [8], and slang words. Twitter is one of the good platforms for code mixed text. Telugu-English tweets were collected from the Twitter API, with a proper authentication process. Annotation is a manual process of labelling sentiment class objects to tweet in the corpus. The sentiment class objects are positive, negative, and neutral. Because of lack of the annotated data in English-Telugu code mixed text, annotation of a text is done with the Telugu native speakers of three professionals.

2.  **Pre-processing of the Collected Corpus of Twitter Data**

Pre-processing of Twitter text is a very essential step in this process. Twitter data contains full of noisy information, like hyper-links, #, @ tags, numbers, misspellings, and slang words. Code mixed data has more possibility of noisy data when compared to the normal text. For better performance, sentiment analysis and noisy data must be nullified. There are two approaches to nullify; one is by removing the symbols, words, tweets, RT, numbers, hyper-links which are not required in the analysis part. And the second one is by replacing the symbols, words, abbreviations, contraction words with proper equal words or extensions.

3.  **Language Identification and Transliteration**

In the code mixed text, language identification [12] is an important task. The language identification is done at the word (aspect) level. The code mixed or code switched text is the combination of more than one language with other linguistic units. In this language identification [7] process, every tweet is tokenized into individual words. Language is identified for each word, for example, the language of a word in English returns as 'en', if it is in Telugu, it is returned as 'te', and if it is in English-Telugu code mixed word, it is returned as 'et'. Every word is embedded with the language tagger.

"mee anna maatalu chudu iragadeesadu"

"monoka tweet ilage vesi delete chesaru veeti grnchi kuda thavvithe"

In the given example, a tweet is tokenized into individual tokens and the language of each word is identified and tagged with words accordingly.

Meell'et' annall'et' maatalull'et' chudull'et' iragadeesadull'et'

Monokall'et' tweetll'en' ilagell'et' vesill'et' deletell'en' chesarull'et' veetill'et' grnchill'et' kudall'et' thavvithell'et'

The words with the language tag as 'et' are the English-Telugu code mixed word. This of type code mixed data is also called a Roman script. Roman script words are not suitable for any text analysis.

Transliteration is the process to convert the Roman script into the original tongue [13]. English-Telugu code mixed roman script is transliterated [14] into the Telugu language by using ITRANS format [15]. All roman script words are transliterated into Telugu's original tongue, by using Google transliteration API which is supported in Python.

మీ||'te' అన్న||'te' మాటలు||'te' చూడు||'te' ఇరగదీసాడు||'te'

మోస్కేక్||'te' tweet||'en' ఇలాగే||'te' వేసి||'te' delete||'en' చేసారు||'te' వీటి||'te' గురించి||'te' కూడా||'te' తవ్రితే||'te'

In the above two examples, each word in the tweet is identified by its language and code mixed English-Telugu words are transliterated into Telugu.

## 4. Sentiment Extraction

Sentiment extraction from the Telugu code mixed data is the main objective of this work. Sentiment classification has three objectives positive, negative, and neutral. Telugu SentiWordNet [3, 5, 9] is a lexicon-based approach to extract sentiments from the Telugu text. The Telugu SentiWordNet is the combination of Telugu WordNet + Sentiments. The Telugu SentiWordNet [16] includes 2136 positives words, 4076 negative words, 359 neutral words, and 1093 ambiguous words.

Aspect-based sentiment extraction is the word level prediction, either the word is positive or negative or neutral. A tweet is tokenizing into words for each word language is identified. If it is an English word, SentiWordNet is used to extract the senti score, and if it is a Telugu word, Telugu SentiWordNet is used to extract the Telugu senti score. Sentence-based sentiment extraction is the sentence level extraction. If the sentence contains n tokens, from each token, sentiments are extracted. These tokens may be either English or Telugu. According to the language, sentiments are extracted. Sentiment scores are calculated as follows.

Senti Score $= P_s - N_s$.
$P_s =$ No. of Positive words/Total Words,
$N_s =$ No. of Negative words/Total Words.

మీ||'te' అన్న||'te' మాటలు||'te' చూడు||'te' ఇరగదీసాడు||'te'

మోస్కేక్||'te' tweet||'en' ఇలాగే||'te' వేసి||'te' delete||'en' చేసారు||'te' వీటి||'te' గురించి||'te' కూడా||'te' తవ్రితే||'te'

The above two sentences were classified as positive and negative.

## Experiments and Results

These Twitter data which are collected randomly from the Twitter API from November 2020 to February 2021 are related to Andhra Pradesh Politics. Multilingual tweets are downloaded for the political tag. Telugu code mixed tweets are a part of the multi-lingual tweets. Code mixed approximately from the political tweets are only to compare the movie reviews.

Figure 8.2 explains the collected tweets and their languages, English-Telugu code mixed tweets along with the different language tweets.

The collected corpus from Twitter is 32,456 tweets. English-Telugu code mixed tweets count is 12,658. Manually, sentiment labelling is assigned as positive, negative, and neutral. The required pre-processing steps are applied to the collected code mixed data. All punctuation symbols, duplicate tweets, RT, tags, extra symbols, and emphasize words are in code mixed data. English and Telugu stop word list and all unnecessary text are eliminated. Some of the text is replaced with suitable text. In Fig. 8.3 the sample input data set considered for the experiment with datacleaning and pre processing of raw data done on the tweets is given below.

The roman script is transliterated into the Telugu language. Language identification for a tweet at the word level is verified according to the language. For English words, sentiments are extracted directly, and for Telugu words, sentiments are extracted by using TeluguSentiWordNet. STCMT is a lexicon-based algorithm

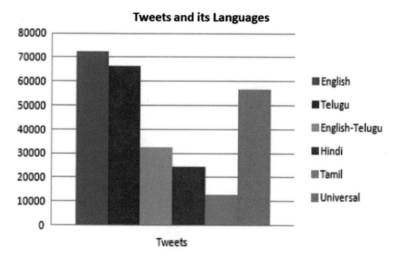

**Fig. 8.2**  Multi-lingual tweets and languages

sno,text,lang
0,evaru chesthunnaru andariki telusu andhuke kada mimmalin,et
1,ide bro problem evaru deep ga chudaru alochincharu keval,et
2,reservations ye kaalam lo vunnav nayana,et
3,mee ayya ni nilabetti year months avtondi elli ah gochi utukkondi ee sari adi kuda undadu antunnaru,et
4,anna mimmalni kalavali plz anna years nunchi try chestunna evvaru help cheyyadam ledu,et
5,sankranthi kalla water vastay rasko ani assembly lo anindhi evaru only ,et
6,em kattaru em kattaru ani aduguthuntey memu anitiki rangulu vesaam ani endhuku cheppuko leka pothunnam a,et
7,years lo rastram lo sontha illu kattkoleni munda evaru,et
8,sare bosu maa gorrelu free money kosam anniyya edo podichesthadu ani c,et
9,mee anna maatalu chudu iragadeesadu,et
10,rahul gandhi valla modi ki eantha advantageoooooooantha kanna double advantage,et
11,bro assembly lo voka magadu challenge,et
12,ninnu padagotte magadu inka puttaledhu mama,et
13,adhi elago istunnaru panilo paniga hostel kuda free aite vere ooru velli chaduvukovachu ani na feeling,et
14,akka manam okka what 's app group create chedam akka ,et
15,kalupu mokkalu anta entento cheppadu,et
16,evaru chesado flag lo paina red colour pettadu,et
17,avunu manam teluguvallam i feel proudly jai jagan anna,et
18,vadunna okkate lekapoina okkate bandi kinda nimmakaya,et
19,free ga istharu emo,et
20,maa amma hospital lo unnapudu billa ajitha nenu exams rasanu aa tharvatha chanipoyindi i have no job no,et
21,shame on this government enti mee nirvakkam rastra hakkulani valdullkkunnaru ennka next entoo,et
22,akkai paruvu poyendhe,et
23,monoka tweet ilage vesi delete chesaru veeti grnchi kuda thavvithe,et
24,veedeni adugu paytm kukka vedu,et

**Fig. 8.3** Sample copy of collected English-Telugu code mixed pre-processed tweets

to extract sentiments from the code mixed data. This is explained in Table 8.1. The accuracy of the STCMT algorithm is 74%, which is compared with the actual class labels and predicted labels. The STCMT algorithm accuracy is also represented in Fig. 8.4.

Figure 8.5 and Table 8.2 represent the machine learning approach about the code mixed of English-Telugu tweets, which is training and testing of 7:3 ratio of code mixed data.

Machine learning approach of sentiment extraction can be done by using different algorithms like support vector machine (SVM), Naive Bayes (NB), decision tree (DT), and finally random forest (RF). The machine learning algorithms are trained with labelled data and predicted with test data set with a ratio of 7:3. Finally, the accuracy of the SVM algorithm is the best prediction for sentiment extraction from English-Telugu code mixed data with up to 86% of accuracy.

**Table 8.1** Lexicon-based sentiment extraction using (STCMT)

| Class | Precision | Recall | F-Score |
|-------|-----------|--------|---------|
| Positive | 0.831 | 0.872 | 0.864 |
| Negative | 0.735 | 0.476 | 0.651 |
| Neutral | 0.450 | 0.681 | 0.621 |

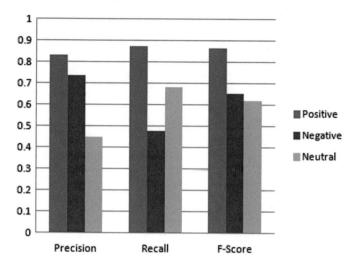

**Fig. 8.4** Lexicon-based sentiment extraction using (STCMT)

**Fig. 8.5** Accuracy of machine learning algorithms training and prediction

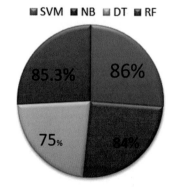

**Table 8.2** Machine learning-based sentiment extraction

| Algorithm | Precision | Recall | F-Score | Accuracy (%) |
|-----------|-----------|--------|---------|--------------|
| SVM | 0.874 | 0.861 | 0.842 | 86 |
| NB | 0.821 | 0.842 | 0.830 | 84 |
| DT | 0.761 | 0.750 | 0.746 | 75 |
| RF | 0.842 | 0.853 | 0.83.8 | 85.3 |

## Conclusion and Future Work

English-Telugu code mixed data is increasing in the open social network like Twitter by the people of Andhra and Telangana states. Sentiment extraction of English-Telugu code mixed data is a task. In the proposed algorithm, Sentiment Extraction from Telugu Code Mixed Tweets (STCMT) is used. All the required steps are to be taken to extract the sentiments, and these pre-processing steps are implemented. Language identification of each word in the tweet is transliterated into Telugu script. By using TeluguSentiWordNet, sentiments are extracted from the Telugu words, and the accuracy is improved up to 74%. From the machine learning approach, training and testing with SVM improve the accuracy up to 86%.

The English-Telugu code mixed tweets do not have the accurate Roman script like a free form of roman script and misspellings that are not transliterating into Telugu script. To improve the sentiment classification, a proper roman script and Telugu is to be maintained. The future work in this research is spelling correction at the level of roman script and converting the free form of the roman script into an accurate roman script by using a machine learning approach.

## References

1. Patra, B., Das, D., Das, A.: Sentiment analysis of code-mixed Indian languages: an overview of SAIL_Code-mixed shared task @ICON-2017 (2018)
2. Sarkar, K.: JU_KS@SAIL_CodeMixed-2017: sentiment analysis for Indian code mixed social media texts (2018). ArXiv, abs/1802.05737
3. Naidu, R., Bharti, S.K., Babu, K.S., Mohapatra, R.K.: Sentiment analysis using Telugu Senti-WordNet. In: 2017 International Conference on Wireless Communications, Signal Processing and Networking (WiSPNET), Chennai, 2017, pp. 666–670. https://doi.org/10.1109/WiSPNET.2017.8299844
4. Das, A., Gambäck, B.: Sentimantics: the conceptual spaces for lexical sentiment polarity representation with contextuality. In: The 3rd Workshop on Computational, Approaches to Subjectivity and Sentiment Analysis (WASSA), ACL 2012, pp. 38–46, Jeju, South Korea
5. Mahata, S.K., Makhija, S., Agnihotri, A., Das, D.: Analyzing code-switching rules for English–Hindi code-mixed text (2020). https://doi.org/10.1007/978-981-13-7403-6_14
6. Mishra, P., Danda, P., Dhakras, P.: Code-mixed sentiment analysis using machine learning and neural network approaches (2018)
7. Das, A., Gambäck, B.: Identifying languages at the word level in code-mixed Indian social media text. In: Proceedings of the 11th International Conference on Natural Language Processing, pp. 378–387 (2014)
8. Malgaonkar, S., Khan, A., Vichare, A.: Mixed bilingual social media analytics: case study: live Twitter data, pp. 1407–1412 (2017). https://doi.org/10.1109/ICACCI.2017.8126037
9. Garapati, A., Bora, N., Balla, H., Sai, M.: SentiPhraseNet: an extended SentiWordNet approach for Telugu sentiment analysis. Int. J. Adv. Res. Ideas Inno. Technol. 5, 433–436 (2019)
10. Das, A., Bandyopadhyay, S.: Dr. Sentiment knows everything!. In: ACL/HLT 2011 Demo Session, pp. 50–55, June, Portland, Oregon, USA
11. Das, A., Bandyopadhyay, S.: Dr. Sentiment creates SentiWordNet (s) for Indian languages involving internet population (2010)
12. Barman, U., Das, A., Wagner, J., Foster, J.: Code mixing: a challenge for language identification in the language of social media. CodeSwitch @EMNLP (2014)

13. Padmaja, S., Bandu, S., Fatima, S.S.: Text processing of Telugu–English code mixed languages (2020). https://doi.org/10.1007/978-3-030-24322-7_19
14. Ahmed, U.Z., Bali, K., Choudhury, M., Sowmya, V.B.: Challenges in designing input method editors for Indian languages: the role of word-origin and context. In: WTIM@IJCNLP (2011)
15. Chopde, A.: Itrans-Indian language transliteration package (2006). http://www.aczone.com/itrans
16. Das, A., Bandyopadhyay, S.: "SentiWordNet for Indian languages. In: The 8th Workshop on Asian Language Resources, pp. 56–63 (2010)

**Arun Kodirekka** a Research scholar at Acharya Nagarjuna University in the Department of Computer Science and Engineering. His areas of interest are Machine learning, Text Analytics and Artificial Intelligence. He has nine years of teaching experience in engineering colleges. He has published eight international journals.

**Dr. Ayyagari Srinagesh** is presently working as a Professor in the Department of Computer Science and Engineering at RVR & JC Engineering College, Guntur. His areas of interest are Image processing, Big Data, Machine Learning, and Artificial Intelligence. He has 21 years of teaching experience and 4 years of Industry experience. He has 26 international journals and 7 International Conference Publications. He is an active member of ISTE, CSI and ACM.

# Chapter 9
# An Integrated Approach for Medical Image Classification Using Potential Shape Signature and Neural Network

**M. Radhika Mani, T. Srikanth, and Ch. Satyanarayana**

## Introduction

One of the widely used application of computer vision and pattern recognitions is medical image processing. Among various applications in medical field, the medical image classification is a much attention gained research field in nowadays. Representing the medical image is a crucial stage for analyzing, classification, and recognition. These algorithms are efficient to classify the normal and abnormal images efficiently [1]. Further, these algorithms are used for detection of various cancers at an early stage. A fuzzy inference system can be used within the adaptive fuzzy neural network [2]. It uses both powerful features, viz., explicit learning capability and powerful learning capability within a single framework. The MRI T1 weighted image classification system [3] uses filtering with diffusion, edge detection. Further, the image is analyzed with morphology. This method uses B-spline fitting model. It classifies the image into six types of tissues. The visual vocabulary [4] is an efficient way to represent the input image in a more prominent way. This will represent and describe the input image with bag of features. It uses support vector machine (SVM) for the classification of images. A J48 decision tree is used for automatic classification of medical images. This framework also uses random forest classifiers [5]. This framework is found to classify brain CT images. This system has described the input image with various texture features and then classified with the random forest classifier. Voxel-based image classification is found to efficient. For this, voxel-based

M. R. Mani (✉)
Department of CSE, Pragati Engineering College, Surampalem, Andhra Pradesh, India

T. Srikanth
Department of ME, Idle Institute of Technology, Kakinada, Andhra Pradesh, India

Ch. Satyanarayana
Department of CSE, JNTUK, Kakinada, Andhra Pradesh, India

ranking system [6] is designed, and the optimal voxel subsets will selected for the classification stage. The classification system uses SVM classifier.

The artificial neural network can be used along with the syntactic and statistical features [7] of the input medical image. It detects various objects in the input image and estimates the corresponding texture features, curvature information, and inclination information. The medical image recognition can be improved with one of the prominent feature called as modality [8]. It is a visual feature and can used to extract the low level features and texture features of the input image. Further, they can be classified with the neural network. This framework is found to be efficient for both gray scale and color images. A synergic deep learning (SDL) framework [9] is developed to handle various issues regarding image modalities and illuminations. It uses convolutional neural network for classification stage to result less synergy error. A three-level feature extraction system [10] can be used for designing efficient medical image classification system. This system can able to extract various pixel, local, and global features into a single feature vector and then reduced with principal component analysis (PCA). Further, SVM and K0-nearest neighbor classifiers are used for the classification process. The G-statistic [11] is computed to measure he log-likelihood ratio of the local binary pattern (LBP) feature. Further, self-organizing map (SOM) and backpropagation neural network (BP) are used for the classification stage. The shape-based methods represent the input image by using an efficient shape signature [12, 13]. These signatures will consider the input image as an object and represents the salient invariant features as their signature. This representation should be further described into a feature vector, and further, the feature vector can be classified with distance measures [14, 15]. From the literature, it is reported that the shape signature-based representation is efficient than other representations of the medical image. So, the present chapter focuses on the development of a novel approach for shape signature for improving the performance of the classification result. The chapter is organized into four sections. This section gives the introduction to the medical image classification, and "Methodology" discusses about the methodology of the proposed approach. "Results and Discussions" consists of various results of the experiments and their discussion; finally, the conclusions are given in "Conclusion."

## Methodology

The object recognition can be performed with various approaches, viz., texture, location, part, and shape-based approaches. The texture-based methods evaluates the texture features, viz., smooth, roughness, etc. of the input object. The location-based methods evaluate the location of the objects. The part-based methods evaluate various parts of the input object viz., stem, blob, etc. The shape-based methods evaluate the shape of the input object. Among various properties of the input object, the shape-based features are unique for each component of the input object. So, the present chapter uses the shape-based object recognition approach for classification of medical images. The proposed algorithm consists of various stages as listed below.

- Contour representation
- EAL sampling
- Design of shape signature
- Description with feature vector
- SOM-based classification.

The shape of the input object components can be represented with either only contour pixels or the complete region pixels. The present chapter focuses on the design of a novel approach with less complexity. So, the present chapter considered the contour pixels for representing the input object in the first stage. In the next stage, the represented contour pixels should be sampled to a finite number of pixels. For this, the present chapter uses equivalent arc length (EAL) sampling technique. The present chapter uses EAL method to sample the contour pixels of the input object to 128 points.

The sampled points should be further represented efficiently for further process. If the representation is failed, then the further stages would not give efficient classification result. So, the present chapter focuses on the design of a novel shape signature for representing the input object in a more efficient manner. The represented shape signature should cover all the details of the input object. Thus, the present chapter considers the input object in a complex plane $(x + iy)$. The signature is constructed within the complex plane. For designing the shape signature, the potential flow around a cylinder with angle of attack is measured. The measurement of shape signature is given in (9.1).

$$\text{Sig} = U \times \left( I \times e^{-i\alpha} + \frac{a^2}{I \times e^{-i\alpha}} \right) \tag{9.1}$$

where

$\alpha$ represents the angle of attack,
$I$ represents the input object.

In the next stage, the Fourier transformation is applied to describe the designed shape signature into an invariant feature vector. The Fourier transformation is found to yield efficient features from the designed shape signature. The derived features are invariant to various transformations viz., translation, rotation scaling, etc. So, the present work uses the Fourier transformation for deriving the Fourier descriptors for designed potential-based shape signature. In the final stage, the invariant feature vector should be given for the classification stage. This stage can consist of various distance measures or neural networks. When compared to the distance-based methods, the neural network-based methods are found to be efficient for performing the classification. Thus, the present work uses neural network-based classification in the final stage of the methodology. There are various types of neural networks

**Fig. 9.1** SOM topology with 6 × 6 dimension

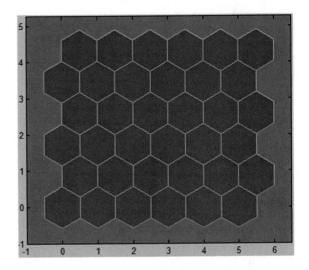

for classification, viz., feed forward neural network, self-organizing map (SOM), learning vector quantization (LVQ), etc. Among these, the SOM is found to be efficient than distance-based methods. So, the present work uses the SOM for handling the classification process with the Fourier descriptors. An example SOM structure with 6 × 6 dimension is shown in Fig. 9.1.

## Results and Discussions

The present work uses contour-based representation for performing the classification. We evaluated the proposed approach on the medical image database [131] with four groups. Each group consists of 131 medical images. The contour-based representation is shown in Fig. 9.2. Fig. 9.2a shows the original medical image, and Fig. 9.2b shows the corresponding contour-based representation. The contour represented is further sampled with equivalent arc length. This gives the result as 128 sampled points. Then proposed signature is designed for the sampled result. For this, the angular attack is considered at 30o. The signature of each group in the database is illustrated in Fig. 9.3. From these results, it is observed that the signature of intragroup images is similar, and the signature of intergroup images is found to be dissimilar. The signature of first group (SC-101) lies in between 0 and 45. Most of the images signature is found in the range of 0–10. The signature of second group (SC-102) is lies in between 0.5 to 4.5. The signature of these images is randomly varying from one to another. The signature of third group (SC-103) lies in between 0 and 2.5. The signature of fourth group (SC-104) lies in between 0 and 7. The represented shape signature is further described with feature vector. For this, the Fourier transformation is used for generating the invariant feature vector. It is identified that first 20 features

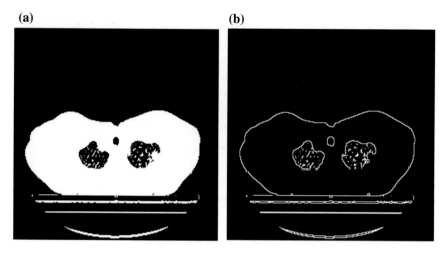

**Fig. 9.2** **a** Original image, **b** contour representation

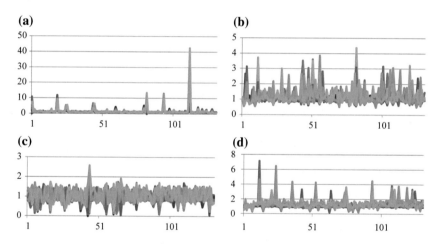

**Fig. 9.3** **a** Shape signature for SC-101, **b** shape signature for SC-102, **c** shape signature for SC-103, **d** shape signature for SC-104

are efficient for the classification. Table 9.1 gives the first nine features of randomly selected nine images.

The description of the input with feature vector is used for performing the classification. The classification process can be performed with the distance measures and neural networks. The neural network is used for the classification process in the proposed approach. Among various neural networks, the current method uses self-organizing map (SOM) for performing the recognition. The evaluated invariant feature description is given to the SOM. The current method evaluates various performance measures of the confusion matrix, viz., false negative ratio (FNR),

**Table 9.1** Sample feature vectors for the input image

| $F_1$ | $F_2$ | $F_3$ | $F_4$ | $F_5$ | $F_6$ | $F_7$ | $F_8$ | $F_9$ |
|---|---|---|---|---|---|---|---|---|
| 35.03152 | 30.4838 | 30.329 | 37.23701 | 40.81016 | 29.12986 | 29.14381 | 29.50082 | 29.49146 |
| 42.90887 | 38.36114 | 38.20634 | 45.11438 | 48.68756 | 37.0072 | 37.02115 | 37.37816 | 37.36879 |
| 43.17701 | 38.62928 | 38.47448 | 45.38252 | 48.9557 | 37.27534 | 37.28929 | 37.6463 | 37.63693 |
| 31.21125 | 26.66353 | 26.50873 | 33.41674 | 36.98988 | 25.3096 | 25.32355 | 25.68056 | 25.67119 |
| 25.02194 | 20.47423 | 20.31943 | 27.22742 | 30.80055 | 19.12029 | 19.13424 | 19.49125 | 19.48189 |
| 45.25409 | 40.70635 | 40.55155 | 47.4596 | 51.03279 | 39.35241 | 39.36636 | 39.72337 | 39.71401 |
| 45.22993 | 40.68219 | 40.52738 | 47.43543 | 51.00862 | 39.32825 | 39.3422 | 39.69921 | 39.68984 |
| 44.61154 | 40.0638 | 39.909 | 46.81704 | 50.39023 | 38.70986 | 38.72381 | 39.08082 | 39.07145 |
| 44.62776 | 40.08002 | 39.92522 | 46.83326 | 50.40645 | 38.72608 | 38.74003 | 39.09704 | 39.08767 |

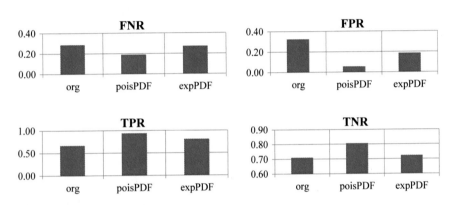

**Fig. 9.4** Comparison of performance measures

true positive ratio (TPR), false positive ratio (FPR), and true negative ratio (TNR). The comparisons are illustrated in Fig. 9.4.

# Conclusion

The shape-based methods are found to be prominent for object recognition. In this, representation of shape of the object is a crucial stage. For this, the present proposes a novel approach for shape signature based on vortex flow on cylinder with angular attack. The proposed signature is further described with Fourier transformation. Finally, SOM-based classification is performed to evaluate the performance measures. It is observed that the Poisson probability distribution function-based approach is yielding efficient results than other approaches.

# References

1. Lashari, S.A., Ibrahim, R.: A framework for medical images classification using soft set. Procedia Technol. **11**, 548–556 (2013)
2. Hosseini, M.S., Zekri, M.: Review of medical image classification using the adaptive neuro-fuzzy inference system. J. Med. Sig. Sens. **2**(1), 49–60 (2012)
3. Shattuck, D.W., Leahy, R.M.: Magnetic resonance image tissue classification using a partial volume model. Neuroimage **13**(5), 856–876 (2001)
4. Caicedo, J.C., Cruz, A., Gonzalez, F.A.: Histopathology image classification using of bag of features and kernel functions. In: Artificial Intelligence in Medicine, pp. 126–135 (2009)
5. Prasad, B.G., Krishna, A.N.: Classification of medical images using data mining techniques. In: Advances in Communication, Network and Computing, pp. 54–59 (2012)
6. Varol, E., Gaonkar, B., Erus, G., Schultz, R., Davatzikos, Ch.: Feature ranking based nested support vector machine ensemble for medical image classification. In: 9th IEEE International Symposium on Biomedical Imaging, Spain (2012)
7. Sharma, N., Ray, A.K., Sharma, S., Shukla, K.K., Pradhan, S., Agarwal, L.M.: Segmentation and Classification of medical images using texture primitive features: application of BAM type artificial neural network. J. Med. Phys. **33**(3), 119–126 (2008)
8. Cramer, K., Jayashree, William, H.: Automatic image modality based classification and annotation to improve medical image retrieval. In: Proceedings of 12th World Congress on Health (Medical) Informatics; Building Sustainable Health Systems
9. Zhang, J., Xia, Y., Wu, Q., Xie, Y.: Classification of medical images and illustrations in the biomedical literature using synergic deep learning. Comput. Vis. Pattern Recogn. (2017)
10. Mueen, A., Sapiyan Baba, M., Zainuddin, R.: Multilevel feature extraction and x-ray image classification. J. Appl. Sci. **7**(8), 1224–1229 (2007)
11. Wang, P., Krishnan, S.M., Kugean, C., Tjoa, M.P.: Classification of endoscopic images based on texture and neural network. In: Proceedings of 23rd International Conference on the Engineering in Medicine and Biology Society (2001)
12. Radhika Mani, M., Varma, G.P.S., Potukuchi, D.M., Satyanarayana, Ch.: A modified shape context method for shape based object retrieval. SPRINGERPLUS **3**(1), 1–12 (2014)
13. Radhika Mani, M., Varma, G.P.S., Potukuchi, D.M., Satyanarayana, Ch.: Design of a novel shape signature by farthest point angle for object recognition. Int. J. Image, Graphics Sig. Process. **7**(1), 35–46 (2015)
14. Radhika Mani, M., Varma, G.P.S., Potukuchi, D.M., Satyanarayana, Ch.: A conformal mapping based shape signature for object recognition. In: Proceedings of the WSEAS 15th International Conference on Applied Computer Science, 20–22 May, 2015, Konya, Turkey, pp. 183–187
15. Radhika Mani, M., Varma, G.P.S., Potukuchi, D.M., Satyanarayana, Ch.: A corner potential flow based shape descriptor for object recognition. In: Advances in Computing, Control and Networking—ACCN 2016, Bangkok, Thailand, 7–8 May, 2016

# Chapter 10
# Brain Tumour Segmentation Using Hybrid Approach

P. Nageswara Reddy, Ch. Satyanarayana, and C. P. V. N. J. Mohan Rao

## Introduction

The segmentation process gives accurate results so that ambiguity can be minimized while recognizing region of interest in human brain. Various conventional and advanced medical segmentation techniques have been developed from past 20 years [1]. The importance of MRI structural analysis of human brain includes the classification of MR images into particular brain tissue types, description of anatomical structures and also the identification of various brain disorders [2]. In classification category, each pixel element in the MR image becomes the type of tissue class, and they are defined in advance. There is interlinking between the classification and segmentation. Segmentation of a given medical image is an important and significant stage for the analysis of medical images. Also, this is the first and most significant stage in most of the medical applications. Especially, for the analysis of the human brain, segmentation is the common tool applied to visualize the anatomical structures of the brain. Analysing changes of pathological regions, surgery planning and image guided interventions require segmentation procedure. In the literature, various types of segmentation approaches with different accuracy have been reported from past three decades. Enormous amount of growth in visualizing brain damage and exploring anatomy of brain provides huge amount of information with an increasingly larger level of quality. Due to huge amount of database, it becomes the tedious and complex problem to analyse these database by the radiologists who have to extract prominent data manually. Sometimes, the manual analysis is prone to error

P. Nageswara Reddy (✉)
Cognizant Technology Solutions India Pvt. Ltd., Chennai, India

Ch. Satyanarayana
Deparment of Computer Science and Engineering, JNTU, Kakinada, AP, India

C. P. V. N. J. Mohan Rao
Avanthi Institute of Science and Technology, JNTUK, Narsipatnam, AP, India

© The Author(s), under exclusive license to Springer Nature Singapore Pte Ltd. 2022    117
Ch. Satyanarayana et al. (eds.), *Machine Learning and Internet of Things for Societal Issues*, Advanced Technologies and Societal Change,
https://doi.org/10.1007/978-981-16-5090-1_10

and often time consuming process. To overcome the issues of brain data analysis, an inventions of computer-based techniques for the betterment of diagnosis and treatment planning are necessary [3]. Hence, the proposed work presents a novel methods for the segmentation of brain tumours and to classify tumour type so that possible diseases can be identified.

## Related Work

The entire success or failure of the application is dependent in the accuracy of the segmentation. Many segmentation techniques are developed for medical image applications. The segmentation method to be used is chosen on the application and the modality of the image used. In recent years, segmentation of the tumour in the MR images of the brain is well thought-out as the active research topic [4]. Active contour models are widely discussed and used in the field of medical image segmentation. These are the contours inside the given image which moves with the influence of forces; internal forces are defined within the curve, whereas external forces are calculated using the data of the image.

Srinivas et al. [5] implemented two segmentation algorithms, namely K-means (KM) clustering and FCM scheme used for partition of MR brain images to locate the tumour. Two algorithms were compared using portioned region, MSE and PSNR as comparative analysis. The experimentation showed that the FCM technique performs better than the K-means algorithm. The FCM scheme consumed the execution time of 8.639 s, whereas K-means algorithm took 22.831 s.

Chen et al. [6] proposed an enhanced FCM scheme to get better precise outputs. Firstly, they have modified the usual regular term used for smoothing by making use of the non-local features which will decrease the consequence of the noise. Secondly, they have used the FCM which in turn uses the distance function which is developed by making use of the few functions which consists the distance and also the covariance which has the preceding probability that can progress the toughness.

At the same time, the unfairness field is developed by using orthogonal foundation functions which will decrease the consequence of concentration inhomogeneity.

Benson et al. [7] mentioned that to diagnose the brain tumour, MRI technique is one of the good methods; they have developed an improvised edition of FCM as well as watershed technique, and in FCM, they have used an efficient scheme for early selection of the centroid which uses histogram computation, and in watershed technique, they have used atlas marker detection technique which they have used to avoid the difficulty of over-segmentation;; three different operations are performed as a pre-processing stage, namely noise elimination, head striping, and contrast enrichment. For improvised FCM clustering, they got accuracy of 0.889 for improvised watershed algorithm, accuracy of 0.93.

Active contour models work on the basis of object boundaries, and features considered with respect to boundaries of images are namely the shape, texture, smoothness, and external as well as internal forces on the desired object. All these factors result the segmentation of desired object boundaries. The closed contours and shapes of the objects in an image can be used to mark the object boundaries.

Niu et al. [8] proposed a new technique for the detection of tumours from MR images for which a region-based active contour models without using re-initialization of the contourhas been uesd. This method is easily executed by applying a simple finite difference approach. To eliminate noise in the image, erosion and thresholding methods were applied. During this, in addition, a contour is initialized in any part of the image, and the interior contours can also be quickly detected automatically.

## Proposed Methodology

The proposed methodology is the combination of FCM and gradient vector flow deformable model. This method is named as fuzzy active contour model [9, 10] and is widely applied for the segmentation of brain tumour. After pre-processing of the MR image, the combination of fuzzy-GVF active contour or hybrid model is used to segment the brain tumour. The resultant image obtained by the FCM–GVF active contour model produces regions that are segmented brain tumours from the given MR image. The block diagram which depicts the hybrid model is shown in Fig. 10.1. Basically, the GVF works on energy minimization principle and the segmentation of

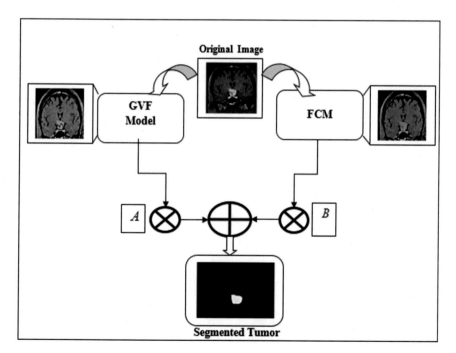

**Fig. 10.1** Segmentation framework of the proposed hybrid model

the object boundary based on the contour evolution. The iteration process controls the contour evolution. The advantage of GVF is that it has the capability to enhance the weak and blurred edges of the object boundaries. This model can also be used for the segmenting the object which has concavity in nature.

## A.    FCM Clustering

The clustering process is an unsupervised method used to develop an algorithm for the distribution clusters. The group of clusters obtained by FCM algorithm are used for minimizing the dissimilar elements in each and every cluster. This approach is popularly applied in the field of medical image processing viz. image enhancement, feature selection, and image segmentation as well as classification. Using FCM method effective classification of pixels in the given image can be grouped into a set of clusters, and these cluster formation, use the principle of Euclidean distance. The FCM builds the fuzzy matrix, and by calculating the Euclidean distance, the different clusters are formed [11]. The following equation can be used for the formulation of the FCM matrix.

$$M = \sum_{m=1}^{s}\sum_{n=1}^{t} S_{mn}^{v} h_{mn}; \quad 1 \leq v \leq \infty \tag{10.1}$$

where $v$ is the fuzziness variable, $h_{mn}$ is the Euclidean distance, $S_{hm}^{v}$ is membership function matrix and also indicates the position of the each pixel, and it denotes the pixels numbers. The distance of the clusters can be expressed as,

$$h_{in} = \|q_i - H_n\| \tag{10.2}$$

where $H_n$ denote the cluster centres and $q_i$ represents the pixel. The cluster centroid considered for initiating the clustering process is computed by the expression given as follows,

$$H_n = \frac{\sum_{m=1}^{s} S_{mn}^{v} q_o}{\sum_{m=1}^{s} S_{mn}^{v}} \tag{10.3}$$

The clustering is carried out by applying $H_n$; the variation of fuzzy matrix can be expressed as,

$$S_{cl} = \frac{1}{\sum_{g=1}^{c} \left(\frac{h_{mn}}{h_{gn}}\right)^{\frac{2}{v-1}}} \tag{10.4}$$

This event continues until the FCM finds an appropriate cluster centroid, and the region of interest in the image represents Mahalanobis distance and depicts the degree of fuzziness.

## B.   Gradient Vector Flow

To overcome the limitation of traditional active contour models, the gradient vector flow model is preferred and makes use of the conventional GVF as the snake's static external force to increase the capture range of the contour as well as its speed of evolution. The potential distance force can be calculated based on model point principle, and the attraction of the contour should be nearest edge points of the object. This method can create difficulties when evolving a contour into concavities of the boundary. To overcome this limitation, a vector diffusion equations which are diffusing the gradients of an edge maps in regions are used. By using equation (10.5) results yielding of different force fields known as gradient vector flow which can be employed in the process of segmenting the desired tumour boundary. The quantity of diffusion varies in accordance with the strength of object edges. This in turn controls the distorting object boundary. The proposed GVF uses conventional energy minimizing expression and it is the beginning point to describe the new snake model called as GVF active contour. This new snake increases the capture range of the contour and also deals with concave boundaries of the object.

$$E_{\text{snake}} = \int_0^1 \alpha |C'(s)|^2 + \beta |C''(s)|^2 + E_{\text{external}} C(s) ds \qquad (10.5)$$

In every case, a binary edge map is necessary but by using the GVF, the continuous gradient space can be directly estimated. In the proposed technique, the first required stage within this framework is for finding the continuous edge map.

The edge map $f(x, y)$ can be extracted from the given image $I(x, y)$.

The fields $\nabla f$ consist of vectors which are pointing towards the edges, but they have limited capture range. In case of homogeneous regions, there are no much information about distinct edges or nearby edges available. In this method, the external force term in Eq. (10.5) is changed with a GVF fields.

$$V(x, y) = [u(x \cdot y) \cdot v(x \cdot y)] \qquad (10.6)$$

Now substituting this external field in Eq. (10.6), the energy functional equation can be written as,

$$E_{\text{GVF}} = \iint \mu \left( u_x^2 + u_y^2 + v_x^2 + v_y^2 \right) + |\nabla f|^2 |V - \nabla f|^2 dx dy \qquad (10.7)$$

where $u_x, u_y, v_x, v_y$ are the spatial derivative values of the fields, and $\mu$ is the blending parameter governing the trade-off between the first term and the second term. On the other hand, when $|\nabla f|$ is large, the second term dominates this function in the object boundary, while the regularization term directs the functions where areas of information is constant.

$$u_t(x, y, t) = \mu \nabla^2 u(x, y, t) - (u(x, y, t) - f_x(x, y)) \cdot \left(f_x(x \cdot y)^2 + f_y(x \cdot y)^2\right)$$
$$\tag{10.8}$$

$$v_t(x, y, t) = \mu \nabla^2 v(x, y, t) - \left(v(x, y, t) - f_y(x.y)\right) \cdot \left(f_x(x \cdot y)^2 + f_y(x \cdot y)^2\right)$$
$$\tag{10.9}$$

It is noticed that these expressions are dismantled, and hence, they can be resolved as separate scalar partial differential equations and are called normalized diffusion equations. The external force used by the GVF makes the capture range of the snake model bigger in size. Because of the GVF usage in the traditional energy formulation, its fundamental principle for diffusing the edge details from the tumour boundary to the remaining part of the image improves. Hence, the combination of FCM and GVF model provides excellent capture range.

## Experimental Results

The proposed method is tested by conducting the experiment with the MRI datasets from brain web database. The image database is available in IBSR web site. It is very confidential about controlling the distribution of medical information of test images. Various databases are available in the registered brain web database. These web sites provide standard test image datasets. Some of the websites are indicated in the reference sections regarding databases. This database consists of several anonymous patients' brain tumour as well as the standard segmentation results called ground truth or bench mark of tumours from these scans.

To evaluate the performance, various performance metrics can be used. The qualitative analysis can also be carried out to visualize the segmentation results. The results are obtained using MATLAB software. For the experimental purpose, sample MR image having tumour history has been used. The following figures illustrate the segmentation process using hybrid approach. The original image shown in Fig. 10.2a is filtered using median filter which removes the artefacts of MR image. After filtering, the contour is initiated in the image as shown in Fig. 10.2c. Once the contour is initialized, it evolves towards the tumour boundary based on the increase in iterations. The positions of the contour for 70 and 150 iterations are shown in Figs. 10.2d, e, respectively. The final tumour boundary extraction by the GVF active contour model is shown in Fig. 10.2f.

The proposed segmentation technique robust and provides segmentation results accurately even in the presence of noise. To test the robustness of the algorithm, the Gaussian noise is added with the original MR image. The results obtained are shown in the Fig. 10.3.

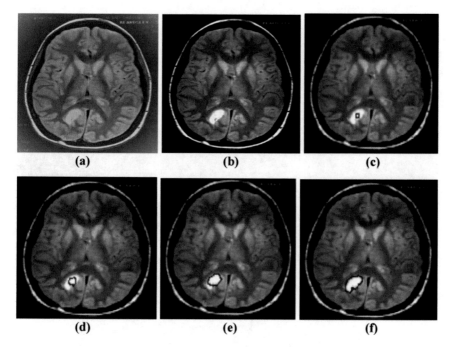

**Fig. 10.2** Segmentation process: **a** original image, **b** filtered image, **c** initial position of the contour, **d** contour evolution for 70 iterations, **e** contour evolution for 150 iterations, and **f** final tumour boundary detection at 220 iterations

## Conclusions

Brain tumour segmentation combing FCM and GVF active contour model called hybrid approach is developed and tested with MRI database. The FCM algorithm classifies pixels of an MR image data set into clusters based on Euclidean distance of a pixel from the centre of the least distant cluster. This helps the contour to converge towards the tumour boundary with a faster rate. As compare to the traditional active contour models, the GVF can detect tumours even in the presence of noise. Because of strong mathematical modelling of GVF active contour models gives the highest segmentation accuracy. The hybrid approach can also be applied for the automatic segmentation techniques. The computational complexity can also be reduced by applying these models.

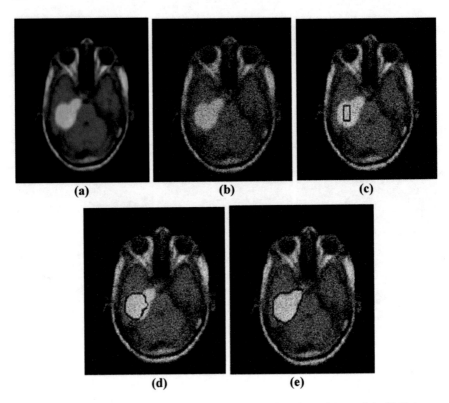

**Fig. 10.3** Segmentation results obtained after adding the Gaussian noise to original MR image: **a** input MR image 2, **b** with Gaussian noise, **c** contour initialization in noisy image, **d** contour evolution for 150 iteration, and **e** final segmented tumour

## References

1. Bauer, S., Wiest, R., Nolte, L.-P., Reyes, M.: A survey of MRI based medical image analysis for brain tumor studies. Phys. Med. Biol. **58**(13), R97 (2013)
2. Paul, T.U., Bandyopadhyay, S.K.: Segmentation of brain tumor from brain MRI images reintroducing K-means with advanced dual localization method. Int. J. Eng. Res. Appl. **5**(16) (2012)
3. Li, M., Xiang, Z., Zhang, L., Lian, Z., Xiao, L.: Robust segmentation of brain MRI images using a novel fuzzy C-means clustering method. In: IEEE 13th International Conference on Natural Computation, Fuzzy Systems and Knowledge. Discoextremely (ICNC-FSKD) (2017)
4. Wadhwa, A., Bhardwaj, A., Singh Verma, V.: A review on brain tumor segmentation of MRI images. Magn. Reson. Imaging **61**, 247–259 (2019). https://doi.org/10.1016/j.mri.2019.05.043,Sep
5. Srinivas, B., Sasibhusana Rao, G.: Unsupervised learning algorithms for MRI brain tumor segmentation. In: IEEE Conference on Signal Processing and Communication Engineering Systems (SPACES), pp.181–184 (2018)
6. Chen, Y., Li, J., Zhang, H., Zheng, Y., Jeon, B., Wu, Q.J.: Non-local-based spatially constrained hierarchical fuzzy C-means method for brain magnetic resonance imaging segmentation. IET Image Proc. **10**(11), 1–35 (2016)

7. Benson, C.C., Deepa, V., Lajish, V.L., Rajamani, K.: Brain tumor segmentation from MR brain images using improved fuzzy C-means clustering and watershed algorithm. In: International Conference on Advances in Computing, Communications and Informatics (ICACCI), Sept. 21–24, Jaipur, India, pp. 187–192 (2016)
8. Niu, S., Chen, Q., de Sisternes, L., Ji, Z., Zhou, Z., Rubin, D.L.: Robust noise region-based active contour model via local similarity factor for image segmentation. Neurocomputing 61(234), 104–119 (2017)
9. Xu, C., Prince, J.L.: Gradient vector flow deformable models. In: Bankman, I. (ed.), Handbook of Medical Imaging. Academic Press, pp. 1–20 (2010)
10. Mekhmoukh, A., Mokrani, K.: Improved fuzzy C-Means based particle swarm optimization (PSO) initialization and outlier rejection with level set methods for MR brain image segmentation, Comput. Methods Programs Biomed. 122(2), 266–281 (2015)
11. Doğanay, E., Kara, S., Özçelik, H.K., Kart, L.: A hybrid lung segmentation algorithm based on histogram-based fuzzy C-means clustering. Comput. Methods Biomech. Biomed. Engi. Imag. Visual. 6(6) (2018)

# Chapter 11
# Identification of Brain Tumors Using Deep Learning Techniques

**T. Jyothirmayi and Ch. Satyananarayana**

## Introduction

Image processing is a vital area in computer science which is gaining much importance due to improvements in technology. Image processing is considered as a method of converting an image into some meaningful form. The applications of image processing include security surveillance, medical imaging, UV imaging, robotics, industrial automation and more. There are many techniques for image processing in which image analysis is having high significance. To analyze an image, the image has to be segmented into different components. Image segmentation is the process of retrieving useful information from images [1]. Segmentation simplifies the image such that it is easy to analyze. Variety of methods has been developed for image segmentation. Threshold method transforms a gray-level image into a binary image. Edge-based method works by finding for most dissimilar pixels which signify discontinuity in image. Region-based methods proceed by searching for similar areas. Model-based method considers image as a collection of image regions.

Image segmentation can be performed in different stages like classification, object detection and finally segmentation. Classification deals with classifying the image into various classes such as people, animals and outdoors. Object detection is used to recognize objects in an image. Segmentation distinguishes portions of a picture and understands what object they have a place with. Fragmentation lays the reason for performing object identification and arrangement.

Object detection is a computer technology related to computer vision and image processing that detects and defines objects. Object detection aims at identifying

T. Jyothirmayi (✉)
Department of Computer Science and Engineering, GITAM University, Visakhapatnam, India
e-mail: jtayi@gitam.edu

Ch. Satyananarayana
Department of Computer Science and Engineering, JNTU Kakinada, Kakinada, India

© The Author(s), under exclusive license to Springer Nature Singapore Pte Ltd. 2022
Ch. Satyanarayana et al. (eds.), *Machine Learning and Internet of Things for Societal Issues*, Advanced Technologies and Societal Change,
https://doi.org/10.1007/978-981-16-5090-1_11

the concepts and locations of objects present in the image. The object detection can also be applied to face detection, pedestrian detection and skeleton detection. Besides classification, object detection provides the additional information which helps in fully understanding the image. Object detection is gaining a lot of importance and is being used in a wide variety of real-time applications. Some of real-time applications of image segmentation using deep learning include recognition of object and face, video surveillance and medical imaging. This chapter focuses on brain tumor detection using image segmentation.

## Literature Study

Lately, different strategies have been proposed for image division, arrangement and identification systems for detecting tumors.

Jain et al. [2] implemented a hybrid approach including discrete wavelet transform, genetic algorithm and RBF neural network for brain tumor classification. The proposed method showed better performance in improving accuracy and minimize the RMS error. Vani et al. [3] presented a prototype for SVM-based object detection for classification of brain MRI images. The proposed method used support vector machine algorithm on structural risk reduction for classification. Besides a Simulink was developed for tumor classification to classify whether classified image is cancerous or non-cancerous. Parveen et al. [4] proposed information digging techniques for the X-ray pictures. The arrangement is made in four stages: preparing, division, highlight withdrawal and characterization. In the principal stage, upgrade and skull stripping is performed to improve the speed and precision. Divisions are finished by using fuzzy C-mean (FCM) grouping. Dark-level run length network is utilized for withdrawal of highlights from mind image; after that, SVM method is applied on the cerebrum brain pictures, giving progressively compelling outcomes for the arrangement of cerebrum MRI pictures.

Akakin et al. [5] proposed the framework for more than one picture inquiry. The highlight is divided into two parts; for color withdrawal, he has used the shading spaces CIELab and tone immersion esteem (HSV) shading space. All the 26 shading and dark scale highlights are extricated utilizing three different shading spaces for the given picture. Surface components are separated utilizing co-event histograms. He has utilized the two different classifiers (SVM and KNN) for the characterization of pictures. Guruvasuki et al. [6] planned a technique utilizing a many bolster vector machine classifier. The picture is processed using a middle channel. The gray-level co-occurrence matrix is utilized for highlight withdrawal. MSVM classifier is utilized for ordering three sorts of pictures. Framework execution can be improved using various picture inquiries than single picture questions.

Mohanapriya et al. [7] proposed a strong recovery utilizing the regulated classifier, which is used to focus on extricated highlights. Dark stage co-event grid calculation can be actualized to separate the surface highlights from the pictures. The element advancement can be done on the separated highlights to choose the best highlights out of it in order to prepare the classifier. The arrangement is performed on a dataset and can be characterized into three classes, for example, ordinary, generous and harmful. They have used the support vector machine (SVM) classifier followed by K-closest neighbor (KNN).

Ramasubramanian et al. [8] structured the multilevel framework for the minuscule pictures that have more than one ailment. The highlights that depend on shading and surface are removed. A recursive SVM classifier with the assistance of removed highlights groups the pictures. In the following level, the comparable pictures are recovered utilizing decision algorithm calculation and accomplish the exactness 96% for FL and 98% for NB.

Baraskar et al. [9] accomplished a comprehensive and deep survey of various brain tumor classification and detection techniques for MRI brain image. A comparative study is made for various techniques. They demonstrated that accuracy, reliability and computational time are the most importance to be considered to compare this technique efficiently, as the diagnosis of brain tumor is a complicated and sensitive task. Kaur et al. [10] analyzed the image processing techniques such as equalized image, feature extraction and histogram equalization which have been developed for extraction of the tumor in the MRI images of the cancer affected patients. Support vector machine (SVM) algorithm works on structural risk minimization to classify the images. The SVM is applied to MRI images for the tumor extraction, and a Simulink model is developed for the tumor classification function.

## Proposed Method

Brain tumor detection is one of the applications in area of medical imaging where image segmentation is playing a key role. Detecting brain tumor is a complex task due to its size, shape, location and type. If tumor is detected in early stages, then there is a high chance of treatment being success for a patient as treatment depends on the stage of tumor diagnosed. The sequential CNN model with dropout mechanism has been trained with the available MRI images dataset for designing tumor detection system.

**Dataset**:

The model is built using the brain MRI Images. The size of datasets used for the model is given in Table 11.1

Below is the sample dataset of brain images shown in Fig. 11.1.

**Table 11.1** Dataset size

| S. No. | Dataset | Size |
|--------|---------|------|
| 1 | Training dataset | 193 |
| 2 | Testing | 10 |
| 3 | Validation | 50 |

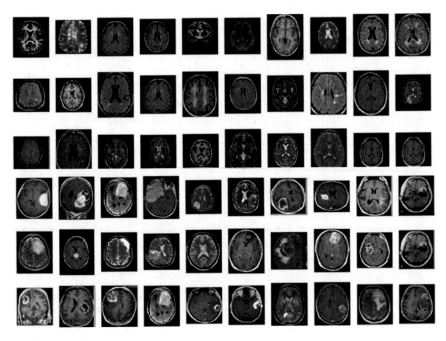

**Fig. 11.1** Sample dataset

On training the system with the dataset, classification of objects can be done. Then the system is provided with test dataset to predict the output along with their respective accuracies. Based on the prediction of output and accuracy, the performance measure of the system is evaluated.

**Flow of the Proposed System** (Fig. 11.2):

**ImageAI**

ImageAI is a Computer Vision Python library to build applications and systems. It provides classes to train and perform image recognition tasks. It also provides approach for training custom object detection models using YOLOv3 architecture. The detection model training class trains the object detection model on image datasets that are in Pascal VOC annotation format using YOLOv3. The training process generates a JSON file that maps the objects names in the image dataset.

**Label Img**

Label Img is an image annotation tool. Annotations are saved in Pascal VOC format as XML files. Pascal Visual Object Classes (VOC) format is a format for providing object detection data, i.e., images with bounding boxes.

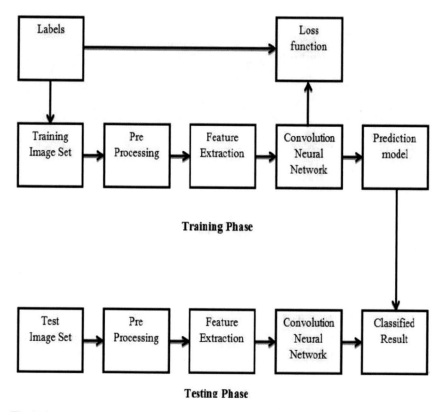

**Fig. 11.2**   Block diagram of proposed model

## Keras

Keras is an open-source neural network library written in Python that runs on top of TensorFlow. It is modular, fast and easy to use. Keras High-Level API handles the models, defines layers or sets up multiple input–output models.

## Results and Discussions

The performance of model built is measured by calculating different metrics like precision, accuracy and recall. Precision is calculated as the correct positive predictions by that of total positive predictions, accuracy is the correct predictions in the set of data divided by total number of values and recall is the total number of true positive values by all positive results summed up. The model proceeds by loading the images from dataset, crops them and performs preprocessing of the image as shown in Fig. 11.3.

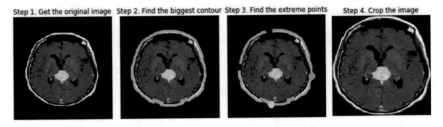

**Fig. 11.3** Preprocessing the image

**Table 11.2** Confusion matrix for validation data

|            | Predicted Yes | Predicted No |
|------------|---------------|--------------|
| Actual Yes | 16            | 3            |
| Actual No  | 2             | 29           |

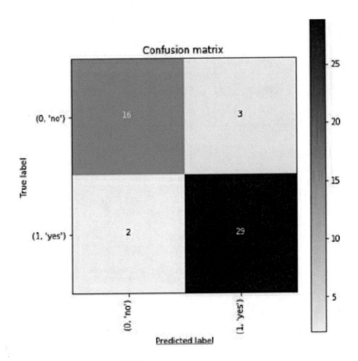

**Fig. 11.4** Confusion matrix for validation

The features are extracted from the preprocessed image, and the model is built using sequential CNN on training dataset. While the model is trained, the confusion matrix generated specifies that model was able to detect tumor T.

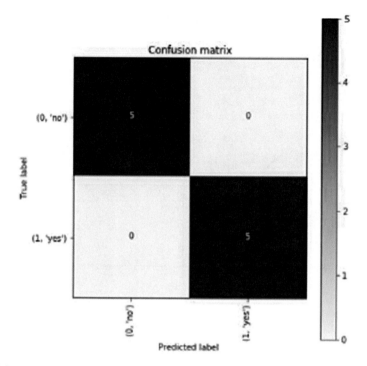

**Fig. 11.5** Confusion matrix for testing data

**Confusion Matrix**: This model has been tested for 50 random images. Based on these images, the confusion matrix generated was shown in Table 11.2 and Figs. 11.4 and 11.5.

The model was accurate with 96% and failed to detect tumor in very minimum number of cases. The performance of model is shown in Table 11.3.

The model accuracy and loss on training and testing data are summarized in Figs. 11.6 and 11.7. It indicates that the model is having better performance.

The proposed method proved to be more accurate when compared with existing models, and it is summarized as follow in Table 11.4 and Fig. 11.8.

**Table 11.3** Performance measures of proposed model

|  | Precision | Recall | Support |
|---|---|---|---|
| No | 0.95 | 0.91 | 25 |
| Yes | 0.94 | 0.93 | 27 |
| Weighted average | 0.94 | 0.92 | 26 |

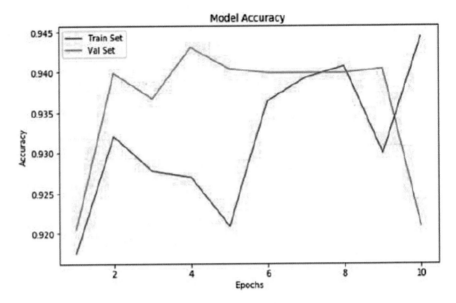

**Fig. 11.6** Model accuracy

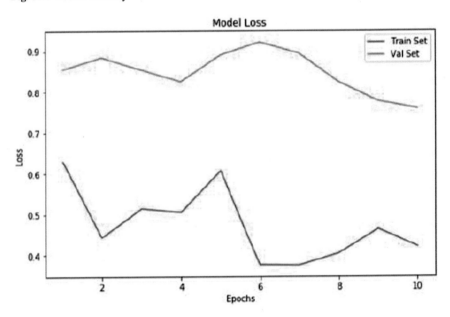

**Fig. 11.7** Model loss

**Table 11.4** Comparison of proposed method with other methods

| Method | SVM | FCM | Random forest | Decision tree | Proposed method (CNN) |
|---|---|---|---|---|---|
| Accuracy in percentage | 82 | 91 | 90 | 95 | 96 |
| Method | SVM | FCM | Random forest | Decision tree | Proposed method (CNN) |
| Accuracy in percentage | 82 | 91 | 90 | 95 | 96 |

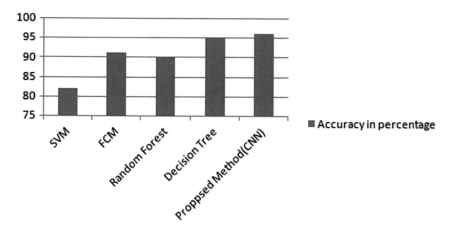

**Fig. 11.8** Comparison of proposed model with other methods

## Conclusion

The proposed model is developed to structure effective programmed mind tumor characterization with high precision, execution and low multifaceted nature. In the regular mind, tumor characterization is performed by utilizing fuzzy C-means [FCM]-based division, surface and shape include extraction and SVM, and decision tree-based and other methods. Yet, the calculation time is more within the meanwhile exactness is low. In order to enhance the exactness and to diminish the calculation time, a convolution neural system-based mostly characterization is conferred within the projected conspiracy. Likewise, the order results are given as a tumor or ordinary cerebrum pictures. CNN is one among the profound understanding of technologies, which consists of classes for forwarding layers. The model performance was measured with accuracy of 96% which was better when compared with other existing models of image segmentation. In future, the model can be further extended by a methodology for fixed scaling which needs further investigation.

# References

1. Gonzalez, R.C., Woods, R.E.: Digital Image Processing, 3rd edn. Prentice-Hall (2009)
2. Naik, J., Patel, S.: Tumor detection and classification using decision tree in brain MRI. Int. J. Eng. Dev. Resear. IJEDR (2013)
3. Vani, N., Sowmya, A., Jayamma, N.: Brain tumor classification using support vector machine. Int. Res. J. Eng. Technol. (IRJET) (2017)
4. Saha, S.: A Comprehensive Guide to Convolutional Neural Networks—the ELI5 Way (2018)
5. Akakin, H.Ç., Gokozan, H.N., Otero, J.J., Gurcan, M.N.: An adaptive algorithm for detection of multiple-type, positively stained nuclei in IHC images with minimal prior information: application to OLIG2 staining gliomas. Med. Imag. Dig. Pathol. 942007 (2015)
6. Guruvasuki, A., Arasi, J.P.: MRI brain image retrieval using multisupport vector machine classifier. Int. J. Adv. Inf. Sci. Technol. **10**(10), 29–36 (2013)
7. Mohanapriya, S., Vadivel, M.: Automatic retrival of MRI brain image using multiqueries system. In: International Conference on Information Communication and Embedded Systems (ICICES). https://doi.org/10.1109/ICICES.2013.6508214
8. Ramasubramanian, B., Prabhakar. G., Murugeswari, S.: A novel approach for content based microscopic image retrieval system using decision tree algorithm. Int. J. Sci. Eng. Res. **4** (2013)
9. Baraskar, S.H., Kamble, B., Survase, A.: An implementation of brain tumor detection and classification using image processing. Int. J. Adv. Sci. Res. Eng. Trends **3**(4) (2018)
10. Kaur, N., Bhati, R.: Development of a faster region based convolution neural network technique for brain image classification. Int. J. Comput. Sci. Eng. **8**(6), 18–24 (2020). https://doi.org/10.26438/ijcse/v8i6.1824

# Chapter 12
# Implementation of ANN for Examining the Electrical Parameters of Cadmium Sulfide Solar Cell

S. V. Katkar, K. G. Kharade, S. K. Kharade, Naresh Babu Muppalaneni, K. Vengatesan, and R. K. Kamat

## Introduction

The development of the total general population is expanding step by step, building energy usage. Because of high energy requests worldwide, the familiar non-renewable energy sources would diminish in 2050. Solar energy is a renewable source of energy. The generation of voltage in a solar cell is known as a photovoltaic effect or photoelectric effect. The photovoltaic effect was identified by Edmund Becquerel in 1839, and he explains how electricity is generated from sunlight [1]. Using coated selenium, first solar cell was created by Charles Fritts in 1883. In 1941, Russell Ohl invented the silicon solar cell in 1941. MATLAB means MATrisLABoratory. By using MATLAB, we create models to analyze the information. It is also useful. The present research focuses on thin-film solar cells in generation second, and these solar cells are available at a minimum cost. Jing Wei et al. reported that there are different

S. V. Katkar · K. G. Kharade (✉)
Department of Computer Science, Shivaji University, Kolhapur, Maharashtra, India
e-mail: kgk_csd@unishivaji.ac.in

S. V. Katkar
e-mail: svk_csd@unishivaji.ac.in

S. K. Kharade
Department of Mathematics, Shivaji University, Kolhapur, Maharashtra, India
e-mail: skk_maths@unishivaji.ac.in

N. B. Muppalaneni
Department of Computer Science and Engineering, NIT Silchar, Silchar, Assam, India

K. Vengatesan
Computer Engineering, Sanjivani College of Engineering, Kopargaon, India

R. K. Kamat
Department of Electronics, Shivaji University, Kolhapur, Maharashtra, India
e-mail: rkk_eln@unishivaji.ac.in

© The Author(s), under exclusive license to Springer Nature Singapore Pte Ltd. 2022
Ch. Satyanarayana et al. (eds.), *Machine Learning and Internet of Things for Societal Issues*, Advanced Technologies and Societal Change,
https://doi.org/10.1007/978-981-16-5090-1_12

traditional approaches are used for finding new material. Old materials are not given optimum solutions for today's problem, so machine learning provides powerful data processing with better performance; it can be done using different tasks like material detection, analysis, and design [2]. Dutt and Nikam studies twenty years of solar cell research data in India, i.e., (1991–2010). Hafezi and Karimi reported that the new thin-film solar cell contains hydrogenated microcrystalline and amorphous silicon layers known as micromorph tandem solar cells. Mahfoud et al. studied solar cells' behavior [3]. Hadjab and Berrah investigated a new method for modeling current and photovoltaic panel voltage using artificial intelligence for modeling of solar cell. They focused on modeling the power, current, and voltage of solar PV and how it changed with temperature, irradiation, etc. This module was trained by using MATLAB and compares the simulated results with experimental results. The error found during the simulation of the module is very less; this confirms the effectiveness of the results [4]. Chunjun Liang et al. observed the polymer solar cells' concert using the bulk heterojunction electrical and optical model. Here, the simulation is done with bulk hetero junction solar cell, which was based on P3HT. The obtained results are compared with experimental data [4]. Dongale et al. have successfully simulated and developed tandem silicon solar cell using an artificial neural network. Here, they use NanoHub for simulation of tandem solar cell by changing the thickness of silicon solar cell layer; they check the efficiency of solar cell. He also developed an artificial neural network for silicon solar cell and compared the predicated result with simulation results. He identified the negligible change in the simulation results and predicated results [5]. Katkar et al. find the solar cell's electrical parameters, i.e., the voltage at the maximum point and current at a full point by implementing a Java-based framework [6]. K.G. Kharade et al. report the effectiveness of CZTS solar cells over other conversion devices. By using MATLAB, he developed a model with three input parameter and one output parameter. Input parameters considered as Jsc, Voc and FF, and efficiency was considered as a output parameter. By changing the number of hidden neuron, he checks the result generated by simulation tool. By changing the number of hidden neuron, they observed the different average errors. At hidden neuron 5, they found approximate result

## Working Model of CdS Thin-Film Solar Cell

Artificial neural network (ANN) is useful for modeling the nonlinearities of CdS thin-film solar cell. For nonlinear problems, it gives the most appropriate solution. Here, we used feed-forward algorithm, linear transfer function, and hyperbolic tangent sigmoid transfer function [7]. In this network, the hidden layers, as well as output layers, are present. The network consists of a hidden layer and a layer of outputs. To train this architecture feed-forward algorithm, Levenberg–Marquardt is used [8] (Fig. 12.1).

Here, the input of the CdS thin-film solar cell contains.

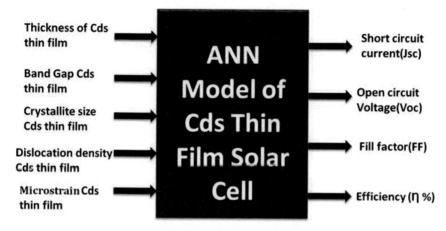

**Fig. 12.1**  Proposed artificial neural network model of CdS

- Thickness,
- Band gap,
- Crystallite size,
- Dislocation density, and
- Microstrain.

The output of the solar cell contains Jsc, Voc, FF and ƞ. Measure the performance in terms of mean square error and correlation coefficient the performance of ANN model of CdS thin film solar cell [9, 10].

## Artificial Neural Network Modeling of CdS Thin-Film Solar Cell

For CdS thin-film solar cell, following data was considered as input for neural network.

- Thickness (nm),
- Band gap (eV),
- Crystallitesize (nm),
- Dislocation density (lines m$^{-2}$),
- Microstrain (lines$^{-2}$)

Table 12.1 show the data input to a network that data comes from an experiment.
Table 12.2 shows the data output that comes from experiments for CdS thin-film solar cell.

**Table 12.1** Input parameters provided for neural network collected from a material scientist

| Name of the solar cell | Thickness (nm) | Band gap (eV) | Crystallite size (nm) | Dislocation density (lines m-2) | Microstrain (lines-2) |
|---|---|---|---|---|---|
| CdS thin-film solar cell | 570 | 2.23 | 61 | 2.68 | 5.88 |
| | 691 | 2.12 | 66 | 2.29 | 5.42 |
| | 722 | 2.05 | 75 | 1.77 | 4.45 |
| | 713 | 1.94 | 86 | 1.35 | 4.12 |

**Table 12.2** Tabular representation of experimental outcomes for CdS

| Name of the solar cell | Experimental results | | | |
|---|---|---|---|---|
| | Jsc | Voc | FF | Efficiency η |
| CdS thin-film solar cell | 0.153 | 253 | 39 | 0.01 |
| | 0.176 | 244 | 45 | 0.02 |
| | 0.347 | 279 | 44 | 0.04 |
| | 0.846 | 293 | 41 | 0.10 |

## Results and Discussion

Table 12.3 depicts the ANN model performance for CdS thin-film solar cell. It also describes the measurement of the correlation coefficient given by the network and expected output [11].

**Table 12.3** Implementation of the system at various numbers of hidden neurons

| No. of hidden neurons | Training percentage (%) | Validation percentage (%) | Testing percentage (%) | Dataset | Correlation coefficient | Average correlation coefficient |
|---|---|---|---|---|---|---|
| 5 | 75 | 10 | 15 | Training | $1.0000\ e^{-0}$ | 0.99965 |
| | | | | Validation | $9.9979\ e^{-1}$ | |
| | | | | Testing | $9.9972\ e^{-1}$ | |
| 10 | 85 | 10 | 5 | Training | $1.0000\ e^{-0}$ | 0.99916 |
| | | | | Validation | $9.9998\ e^{-1}$ | |
| | | | | Testing | $9.9979\ e^{-1}$ | |
| 15 | 80 | 15 | 5 | Training | $1.0000\ e^{-0}$ | 0.99966 |
| | | | | Validation | $9.9992\ e^{-1}$ | |
| | | | | Testing | $9.9966\ e^{-1}$ | |
| 20 | 80 | 10 | 10 | Training | $1.0000\ e^{-0}$ | 0.99925 |
| | | | | Validation | $9.9999e^{-1}$ | |
| | | | | Testing | $9.9966e^{-1}$ | |

**Table 12.4** Predicted result using ANN at hidden neuron value '20'

| Name of the solar cell | Predicted results using ANN | | | |
|---|---|---|---|---|
| | Jsc | Voc | FF | Efficiency ($\eta$) |
| CdS thin-film solar cell | 0.1530 | 253 | 39 | 0.01 |
| | 0.1760 | 244 | 45 | 0.02 |
| | 0.8875 | 272.9226 | 45.7597 | 0.0853 |
| | 0.5309 | 310.0813 | 44.7193 | 0.0706 |

Highlighted outcomes are as follows:

1. It is observed that the all-inclusive correlation coefficient is 0.99965 if we provide '5' number of hidden neurons for the calculation of the training dataset by considering that the correlation coefficient is 1 at the time of the validation dataset, which is 0.99979 and a testing dataset 0.99972.
2. It is observed that the all-inclusive correlation coefficient is 0.99916 if we provide '10' number of hidden neurons for calculation of training dataset by considering correlation coefficient is 1 at the time of validation dataset is 0.9998 and the testing dataset is 0.99979
3. It is observed that the all-inclusive correlation coefficient is 0.99947 if we provide '15' number of hidden neurons for the calculation of the training dataset by considering that the correlation coefficient is 1 at the time of the validation dataset, which is 0.99992 and a testing dataset 0.99966.
4. It is observed that the all-inclusive correlation coefficient is 0.99925 if. We provide '20' number of hidden neurons to calculate the training dataset by considering that the correlation coefficient is 1 at the time of the validation dataset, which is 0.99999 and a testing dataset of 0.99966.

Here, we got very negligible values of MSE and mu (). The smallest values means that our model works effectively. The value of hidden neuron '20' gives the most effective result of CdS thin-film solar cell [12] (Table 12.4).

The following graph noted that at '20' number of hidden neurons; the optimized results of CdS solar cell were found [13].

Figure 12.2 shows the mean square error for the hidden neuron 20. Here, best validation performance was found at epoch 4.

Figure 12.3 represents the gradient value, i.e., $1.4338 \, e^{-0.12}$ at epoch 4 mu value, we found $1 \, e^{-007}$ at epoch 4, validation checks parameter is zero at epoch 4.

Figure 12.4 shows the correlation coefficient; here, correlation coefficient for training dataset is 1, for validation 0.99999, for testing 0.99996, and overall correlation coefficient is 0.99925 at 20 number hidden neuron. Figures 12.5, 12.6 and 12.7 show various structures and weights of neural network with 20 hidden neurons.

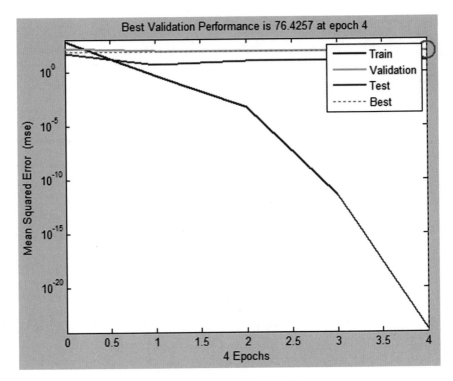

**Fig. 12.2** MSE for the network provided at the time of '20' hidden neurons

## *Impact of Hidden Neurons: Modeling of CdS Solar Cell*

We found the average error of CdS thin-film solar cell data between −1.66727 and 0.962769 for different hidden neurons as shown in Fig. 12.8. The value of hidden neuron '10' represents minimum error [14].

As mentioned in the above graph, we found negative average values of CdS thin-film solar cell where the average value was −1.66727 at the time of 5 no. of hidden neurons, and it was −0.23534 at time of 10 no. of hidden neurons, and it was −1.04528 at time of 20 no. of hidden neurons [9, 15]. The average value becomes 0.962769, which is a positive value at the time of 15 hidden neurons. Figure 12.8 shows that the minimum error found at '5' hidden neurons, so for optimizing solar cell parameters, these outcomes are helpful.

## Conclusion

This chapter aimed to analyze the CdS thin-film solar cell prediction data system using artificial neural network techniques. A CdS performance thin-film solar cell is

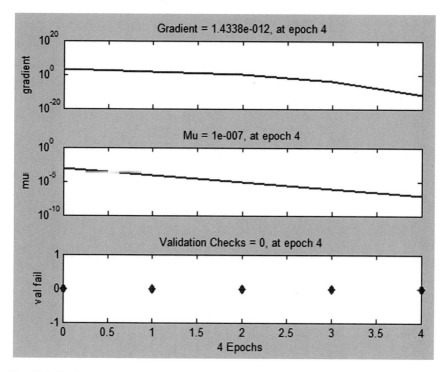

**Fig. 12.3** Gradient, μ and validation parameters of a neural network at hidden neuron '20'

influenced by several factors, such as JSc, Voc, and FF. This chapter analyzes many studies using ANN methods to estimate CdS thin-film solar cells efficiency using the ANN instrument. The results show that the MATLAB instrument is most suitable for predicting performance without laboratory experiments. The outcome obtained by this method is the same as the actual laboratory results. This study successfully simulated CdS thin-film solar cell and modeled its properties using the artificial neural network (ANN). The reported ANN architecture successfully models the nonlinear characteristics of the CdS thin-film solar cells. The present ANN architecture is useful for the development of software for predicting solar cell properties. Finally, the results observed from the trained neural network are in good agreement.

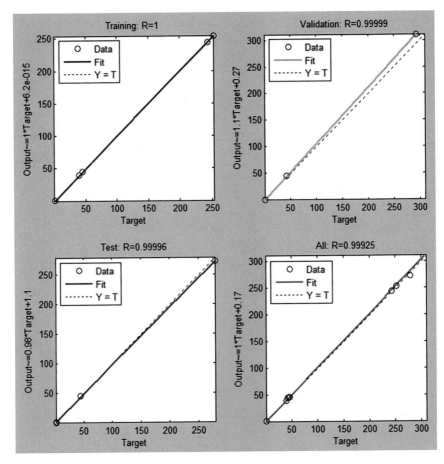

**Fig. 12.4** Correlation coefficient an outcome for the given network at hidden neuron '20'

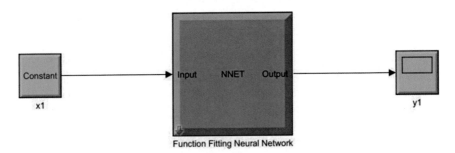

**Fig. 12.5** Input–output mapping in the form of a Simulink diagram for the CdS thin film at the time of 20 hidden neurons

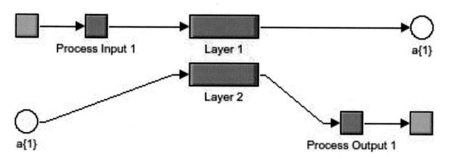

**Fig. 12.6** Interrelation between the hidden layer and an output layer of CdS thin film at the time of '20' hidden neurons

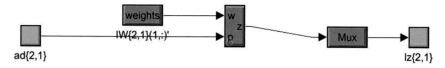

**Fig. 12.7** Weight associated with hidden layer 2 by considering 20 hidden neurons CdS thin film

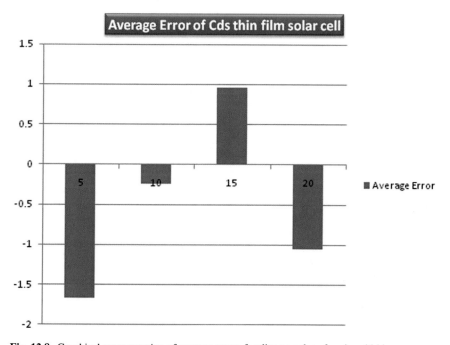

**Fig. 12.8** Graphical representation of average errors for diverse values for given hidden neurons

# References

1. Suzuki, K., Kuroiwa, Y., Takami, S., Kubo, M., Miyamoto, A.: Combinatorial computational chemistry approach to the design of cathode materials for a lithium secondary battery. Appl. Surface Sci. **189**(3–4), 313–318 (2002)
2. Dutt, B., Nikam, K.: Solar cell research in India: a scientometric profile. Ann. Libr. Inf. Stud. 115–127 (2013)
3. Hafezi, R., Karimi, S., Jamalzae, S., Jabbari, M.: Material and solar cell research in high efficiency micromorph tandem solar cell. Ciencia Natura **37**(6–2), 434–440 (2015)
4. Hadjab, M., Berrah, S., Hamza, A.: Neural network for modeling solar panel. Int. J. Energy **6**(1), 9–16 (2012)
5. Katkar, S.V., Kharade, K.G., Kharade, S.K., Kamat, R.K.: An intelligent way of modeling and simulation of WO3 for supercapacitor. In: Thapa, N. (ed.) Recent Studies in Mathematics and Computer Science, pp. 109–117. Book Publisher International, Hooghly, West Bengal, India (2020)
6. Kharade, S.K., Kamat, R.K., Kharade, K.G.: Simulation of dye synthesized solar cell using artificial neural network. Int. J. Eng. Adv. Technol. (IJEAT) 1316–1322 (2019)
7. Mohan, K.R.: Modeling and simulation of solar PV cell with Matlab/Simulink and its experimental verification. J. Innovative Res. Solutions 118–124 (2014)
8. Kharade, K.G., Kharade, S.K., Katkar, S.V., Kamat, R.K.: Selection of small index to reduce the number of pages for improving efficiency. In: Rodino, L.G. (ed.) Recent Studies in Mathematics and Computer Science, vol. 4, Chapter 5, pp. 66–71 (2020)
9. Kharade, K.G., Kharade, S.K., Katkar, S.V., Kamat, R.K.: Performance improvement using covering index to reduce row identifier lookups. In: Rafatullah, M. (ed.) Recent Advances in Science and Technology Research, vol. 3, Chapter 11, pp. 114–119 (2020)
10. Mahfoud, A., Mekhilef, S., Djahli, F.: Effect of temperature on the GaInP/GaAs tandem solar cell performances. Int. J. Renew. Energy Res. (IJRER)**5**(2), 629–634 (2015)
11. Koyama, M.T.: Combinatorial computational chemistry approach for materials design: applications in DeNOx catalysis, fischer-tropsch synthesis, lanthanoid complex, and lithium ion secondary battery. J. Chem. Inform. (2007)
12. Green M.A.: Thin-film solar cells: review of materials, technologies and commercial status. J. Mater. Sci. Mater. Electron. **18**(1), 15–19 (2007)
13. Kubo, M.K.: Combinatorial computational chemistry approach to the high-throughput screening of metal sulfide catalysts for CO hydrogenation process. Energy Fuels 857–861 (2003)
14. Parmar, H.: Artificial neural network based modelling of photovoltaic system. Int. J. Latest Trends Eng. Technol. 50–59 (2015)
15. Kharade, S.K., Kamat, R.K., Kharade, K.G.: Artificial neural network modeling of MoS2 supercapacitor for predicative synthesis. Int. J. Innovative Technol. Exploring Eng. 554–560 (2019)
16. Liang, C., Wang, Y., Li, D., Ji, X., Zhang, F., He, Z.:. Modeling and simulation of bulk heterojunction polymer solar cells. Solar Energy Mater. Solar Cells **127**, 67–86 (2014)
17. Khodakarimi, S., Hekmatshoar, M.H., Abbasi, F.: Monte Carlo simulation of transport coefficient in organic solar cells. Appl. Phys. A **122**(2), 140 (2016)
18. Salunkhe, M.M., Khot, K.V., Sahare, S.H., Bhosale, P.N., Bhave, T.: Low temperature and controlled synthesis of $Bi_2$ $(S_{1-x}$ $Se_x)_3$ thin films using a simple chemical route: effect of bath composition. RSC Adv. **5**(70), 57090–57100
19. Kharade, K.G., Kharade, S.K., Kumbhar, V.S.: Impact of digital India on various sectors. Indian J. Innov. Manage. Excellence Res. (IJIMER), 37–40 (2018)
20. Kharade, K.G., Mudholkar, R.R., Kamat, R.K., Kharade, S.K.: Perovskite solar cell simulation using artificial neural network. Int. J. Emerg. Technol. Innov. Res. 336–340 (2019)

# Chapter 13
# Deduplication of IoT Data in Cloud Storage

**Ch. Prathima, Naresh Babu Muppalaneni, and K. G. Kharade**

## Introduction

Distributed computing has lately risen as a very much preferred plan for utility of IoT data. The possibility of cloud is to create processing assets as an utility or an administration on interest to clients over the sensor data. The prospect of distributed computing is kind of the equivalent as matrix registering, which plans to accomplish virtualization of IoT data [1]. In frame computing, the associations sharing their computing assets, similar to processors, in order to accomplish the most processing ability, while distributed computing intends to deliver processing assets as an utility on interest, which may extent to different clients. This makes distributed computing assume a genuine job inside the business space, though grid framework is very much loved in education, science, and engineering [2]. Many meanings of distributed computing are outlined, relied on the individual reason for read or innovation utilized for framework improvement. We trend to plot distributed computing as a plan of action that give processing assets as an administration on interest to clients over the sensor data [3].

Cloud providers pool figuring assets along serve clients by means of a multi-occupant sensor data Fig. 13.1. Registering assets are conveyed over the IoT wherever clients will get to them through various customer stages. Clients will get to the assets on request whenever while no human cooperation with the cloud provider. From a

Ch. Prathima (✉)
Department of IT, Sree Vidyanikethan Engineering College (Autonomous), Tirupathi, India
e-mail: prathima.ch@vidyanikethan.edu

N. B. Muppalaneni
Department of CSE, National Institute of Technology Silchar, Silchar, India

K. G. Kharade
Department of Computer Science, Shivaji University, Kolhapur, Maharashtra, India
e-mail: kgk_csd@unishivaji.ac.in

© The Author(s), under exclusive license to Springer Nature Singapore Pte Ltd. 2022     147
Ch. Satyanarayana et al. (eds.), *Machine Learning and Internet of Things for Societal Issues*, Advanced Technologies and Societal Change,
https://doi.org/10.1007/978-981-16-5090-1_13

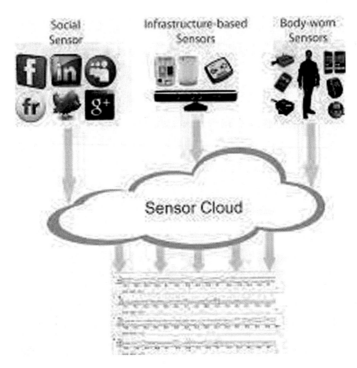

**Fig. 13.1** Enormous data collected from sensors and stored in cloud

client's motivation, registering assets are huge, and customer requests on sensor data are adjusted to fulfill business goals. This is frequently assisted by the adaptability by cloud administrations to scale assets all over on interest contributing the office of virtualization. Also, cloud providers can screen and administrate the use of sensor data for each customer for charge capacities, enhancement assets, and ability to plan and diverse undertakings.

Cloud storage is one among the administrations in distributed computing that gives virtualized capacity on request to clients. Cloud storage might be used in numerous different ways that [4]. For instance, clients will utilize IoT storage as a reinforcement benefit, as against keeping up their very own cache. Associations will move their sensor data to the cloud Fig. 13.2 that they will achieve extra ability at the moderate cost, rather than looking for further physical capacity. Applications running within the cloud conjointly require change or changeless sensor data store in order to help the client applications.

As the quantity of IoT data inside the cloud is progressively expanding, clients hope to succeed in requesting cloud usage whenever, though providers whereas suppliers are needed to take care of system handiness and method an outsized quantity of sensor data. Suppliers want some way to dramatically cut back knowledge volumes, in order that they will cut back prices whereas to spare vitality utilization

**Fig. 13.2** Sensor data transferred to cloud

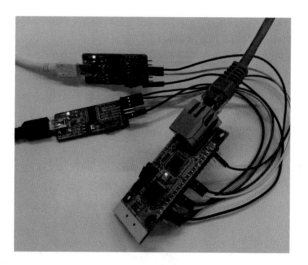

for running huge capacity frameworks. The equivalent as various caches, cache in cloud situations may likewise utilize learning deduplication strategy.

Deduplication might be a procedure whose goal is to help cache power. With the intend to increase sensor data, in ancient deduplication frameworks, copied learning pieces establish and store only one proliferation of the data. Deduplication back each space for storing and network information measure [5]. However, such methods may end up with a negative effect on framework adaptation to internal failure. Thanks to this drawback, several approaches and techniques are planned that not solely give solutions to attain storage potency, however conjointly to boost and help its adaptation to non-critical failure. These procedures give excess of sensor data block after deduplication process.

However, current knowledge deduplications in Fig. 13.3 instruments in distributed storage are static plans connected are applied to all or any knowledge eventualities. For instance, IoT data use in cloud changes in the course of due, some data blocks

Deduplication reduces the amout of stored data

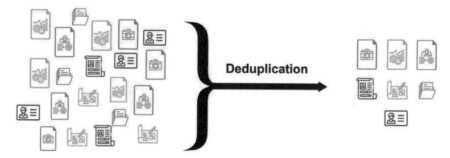

**Fig. 13.3** Deduplication reducing the storage

could likewise be used at a time, anyway probably would not be used in another period of time. On account of setbacks in static plans, that cannot address consistently changing client conduct, deduplication in cloud cache needs a dynamic idea that has the adaptability to adjust to designs and regularly changing client conduct on sensor data in cloud cache.

The contributions of this chapter might be a dynamic learning deduplication subject for distributed storage for IoT data, in order to satisfy a harmony between capacity intensity and adaptation to internal failure necessities, and conjointly to support execution in cloud storage frameworks.

## Background and Related Work

A.   *Deduplication in Cloud storage:*

Deduplication might be a system to downsize space for IoT data. By characteristic excess knowledge, victimization hash esteems to coordinate sensor blocks only one, and making legitimate tips to various duplicates instead of putting away extraordinary genuine duplicates of the sensor information [6, 7]. Deduplication cut backs sensor data volume in this way circle space and system data volume might be decreased that lessen costs. Deduplication might be connected at almost each purpose that sensor data are hold on or sent in cloud storage. Few cloud providers supply no recovery in IoT data [5] and deduplication will not recovery of disaster more functional by duplicating data block once deduplication for increasing mirroring time and data measure savings. Reinforcement and safe caching in clouds may apply facts deduplication in order to reduce physical ability and system movement [8, 9]. Additionally, in movement strategy, we get to send an outsized IoT data of copied picture learning [10]. There are three significant measurements of relocation to consider: knowledge completely transferred, movement time and fix period. Longer movement time and period would be because resources fail. Hence, deduplication will aid movement [11]. To boot, Mandagere et al. express that deduplication calculations replicate the execution of deduplicated sensor data is an issue.

B.   *Problems of Reliability in IoT data*

Acting as deduplication, fewer bits of IoT data are much more important than others. Earlier deduplication approaches do not actualize repetition of sensor data blocks. Subsequently, the sensor blocks should be recreated over the less vital blocks in order to support the architecture. The creators in [12] consider the outcomes of deduplication on the obligation of the cloud-based IoT. They arranged partner degree to deal with boost responsibility by building up a method to mass and experience the significance of every block by inspecting to measure the sensor data blocks that share the same block and utilize this mass to recognize the degree of repetition required for guarantee quality of service.

## C. *Work Done*

Application aware source dedupe [13], causality-based dedupe, and scalable hybrid hash cluster [14]. The majority of existing outcomes that utilize deduplication innovation specialize in spending significant time in the decrease of reinforcement time while overlooking the reclamation time in IoT data. The creators arranged causality-based dedupe, an execution supporter for each cloud reinforcement and cloud reestablish activities that might be a middleware that is symmetrical and might be coordinated into any current reinforcement framework in sensor data. The fundamental point of those associated works is accompanying: surface-to-air missile aims to attain associate degree best exchange off between deduplication strength and deduplication overhead, causality-based dedupe diminishes every reinforcement time and reclamation time. Application aware source dedupe [13] expects to scale back the procedure, increment outturn.

Droplet [15], a shared deduplication cache framework, intended for good outcomes and measurability. It comprises of three segments: one meta server that screens the total sensor data, numerous process servers that run deduplication on computer file stream and different cache junction that store sensor data and deduplicated sensor data chunks.

# Proposed System Model

### Overall Systematic design

Our framework is presently supported customer side deduplication victimization entire record hashing on IoT data. Hashing method is executed at the customer and interfaces with anyone of deduplicators with regards to their time and IoT data. The deduplicator at that point recognizes the duplication by examination with the prevailing hash index in referential server. In early deduplication frameworks, if it is a substitution of hash index, it will be recorded in referential server, and furthermore, the IoT data will be transferred to file servers, and its intelligent way will record in referential server. If it is present, the measure of index for the record will be exaggerated.

Few frameworks might be a scope of duplicates of each record of sensor data with a static number. If the sensor data with an outsized scope of indexing might need addition framework in order to boost handiness. To overcome this issue, some current works brought dimension of repetition into deduplication frameworks. In any case, characteristic dimension of repetition by range of indexing might be a poor measure as a result of files with less indexing could be essential IoT data.

To boost handiness whereas maintaining IoT data, we have a tendency to implement a deduplication framework that thinks about each the changing and using quality of service of the cloud storage. In our framework, when duplication id identified, the redundancy manager at that point figures associate degree best range of duplicates for the record IoT data supported range of references and dimension of quality of service is important. The ranges of duplicates are modified on the regularly changing sensor data, dimension of quality of service are provided for the sensor data records. The progressions are observed, for instance, when a document is erased by a client, or the degree of quality of service of the record has been refreshed, this may trigger the repetition administrator to re-compute associate the best range of duplicates on the IoT data.

Our arranged frameworks demonstrate is appeared in Fig. 13.4. The frameworks consist of the subsequent parts:

**Balancing of Load**: once hashing strategy with Secure Hash Algorithm 1, buyers send a unique Thumb impression (hash esteem) to a deduplicator by means of the balancing of sensor data. The balancing of load takes demands from purchases causing to anybody of deduplicators in keeping their loads at that instant.

**Referential Deduplicators**: An element intended for characteristic the referral duplication by examination with the prevailing hash esteem hold on in referential server. Distributed storage: An information server to store sensor data and assortment of file servers to store genuine IoT data records and their duplicates.

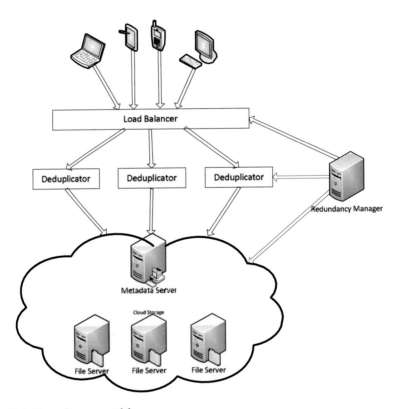

**Fig. 13.4** Planned system model

**Replication Manager**: An element to recognize the underlying scope of duplicates of the IoT data and screen the consistently changing dimension of quality of service.

## Results After Experiments Done

Performed tests on the imitations of our planned model. The tests are conducted for 01, 05, and 10 deduplicators.

Every IoT data used in the tests are made with random substance and properties. There are various sizes of sensor records used in this test: one hundred PC memory unit, 150, 200, 250, 300, 500, 800 Kilobytes, 1,2 Megabytes. Uploading, updating and deleting occasions on 10 documents, 100 records, cardinal records, and 10,000 documents of IoT data. For testing the regularly changing dimension of quality of service, each document has been discretionarily distributed its dimension of quality of services (one-five). One quality of service worth of one-five demonstrates the degree of repetition of each sensor document. Records with top-notch of quality of service will be repeated over the lower sensor data.

When isolated deduplicator is utilized, the framework expandability measurability issues take an all-inclusive time once the measure of records exceeds as appeared in Fig. 13.5. This is frequently because of underneath the critical load with extra demands and extra clients, isolated deduplicator cannot keep up the execution of the IoT data framework. When the measure of deduplicators is exceeds to 5 and 10, the outcomes demonstrate that it scale back the range.

For uploading, every sensor data are transferred to the framework initially, and the amount of time taken is observed on 5 folds deduplicator and 10 folds deduplicators. Including extra deduplicators once the uploading IoT documents increment may facilitate to reduce execution time. The test results in Table 13.1. When the quantities of exchange sensor records are 10 and 100 IoT data documents, victimization 5 deduplicators will scale up to 85.76% and 94.25% of the executed taken by isolated deduplicator, though 10 deduplicators will spare longer at 90.86% and 97.57%. When the measure of exchange IoT data documents has been extended to cardinal records, 5 and 10 deduplicators will facilitate to scale back the interval; however, they are small to 91.45% and 95.60% severally. Time efficient impressively lowers once the uploading IoT data records are higher to 10,000 thousands as 5 and 10 deduplicators will reduce 60.15% and 79.76% of interval.

When IoT data records have been transferred to the framework, we tend to perform tests for the case once there's a consistently changing dimension of quality of service, which implies amount of duplicates of sensor documents within the framework likely could be adjusted with regards to range of quality of service. The consequences of updating IoT data records demonstrate that once the measure of sensor data documents increment, adding extra deduplicators will facilitate to scale the interval. When the quantities of records are 10, 100 documents, 1000 and 10,000 documents, victimization 5 deduplicators will reduce 41.77% and 61.78%, 63.79%, and 75.27% of the

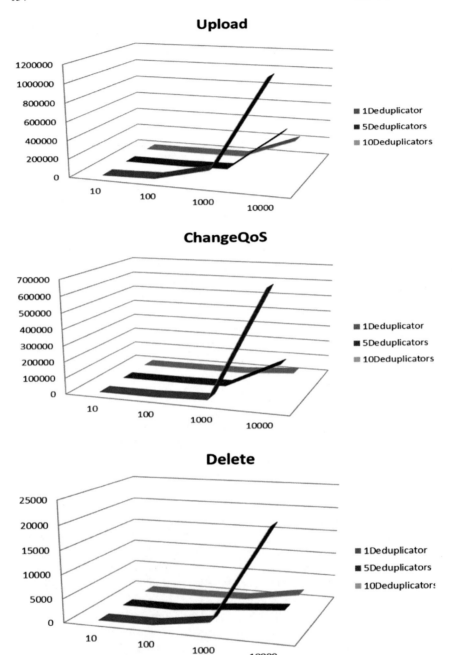

**Fig. 13.5** Experimental results

**Table 13.1** Five and ten duplications percentage of time saving

| Quantity of files | Uploading of data | | Updating of data | | Deletion of data | |
|---|---|---|---|---|---|---|
| | 5 | 10 | 5 | 10 | 5 | 10 |
| 10 | 85.76 | 90.86 | 41.77 | 75.02 | 93.44 | 98.68 |
| 100 | 94.26 | 97.57 | 61.78 | 75.34 | 69.33 | 90.58 |
| 1000 | 91.45 | 95.60 | 63.79 | 82.09 | 40.77 | 85.88 |
| 10,000 | 60.15 | 79.76 | 75.27 | 96.17 | 90.29 | 90.05 |

execution process by Isolated deduplicator, though 10 deduplicators will last longer at 75.02%, 75.34%, 82.09%, and 96.17%. We tend to establish that, when the quantities of IoT data records are 10, 100 and 1000, efficient by including extra deduplicators is nevertheless efficient for the uploading cases. When the quantities of IoT data documents are exaggerated to something like 1000 and 10,000 sensor data records, the efficiency of 5 and 10 deduplicators still increment, in distinction to the uploading cases on the sensor data.

We perform analyses for deleting of IoT data records. Adding extra deduplicators may likewise reduce execution interval; anyway, the consequences of deletion sensor data documents are marginally totally unique in relation to the updating and uploading cases. The outcomes demonstrate that once the quantities of records are 10, 100 documents, 1000 and 10,000 records, exploitation 5 deduplicators will decrease 93.44% and 69.33%, 40.77%, and 90.29% of the intervals taken by isolate deduplicator, though 10 deduplicators will spare longer at 98.68%, 90.58%, 85.88%, and 90.05%. We can see that, for the deleting cases, efficient by including extra deduplicators is small once the quantities of IoT data documents are exaggerated from 10 to 100 and 1000 sensor data records. Nonetheless, when the quantities of IoT data records are exaggerated to 10 thousands, extra deduplicators facilitate to expand efficient.

The test outcomes do not seem to be essentially shocking. Adding additional deduplicators will facilitate to scale back the interval. However, we have a tendency to still establish what are the best range deduplicators to be valuable into the framework with regards to the situations and furthermore the range of IoT data records by then. Also, the outcomes to be assessed against constant data collected through sensors in IoT.

## Conclusion

Distributed storage administrations gave distributed computing has been expanding in quality. It offers on interest virtualized capacity assets and clients exclusively get the volume of IoT data that they really wanted. Since the expanding request and IoT data store within the cloud, facts deduplication is one among the procedures to enhance caching potency. In any case, current facts deduplication procedures in distributed caching are constant topic that restrains their full persistence in effective sensor data in distributed caches.

In this chapter, we tend to propose an effective IoT data deduplication subject for distributed caching, in order to meet a harmony between regularly changing capacity intensity and adaptation to non-critical failure necessities, and conjointly to boost execution in distributed caching frameworks. We tend to dynamically modify the amount of duplicates of sensor data documents with regards to the regularly changing dimension of quality of service. The test outcomes demonstrate that our arranged framework is acting admirably and may deal with measurability drawbacks. We tend to conjointly decide to screen the consistently changing of clients' interest of sensor data documents. Additionally, we tend to choose to assess availability and execution of the framework on the IoT data.

# References

1. Qabil, S., Waheed, U., Awan, S.M., Mansoor, Y., Khan, M.A.: A survey on emerging integration of cloud computing and internet of things. Int. Conf. Inf. Sci. Commun. Technol. (ICISCT) **2019**, 1–7 (2019). https://doi.org/10.1109/CISCT.2019.8777438
2. Dillon, T., Chen, W., Chang, E.: Cloud computing: issues and challenges. In: 2010 24th IEEE International Conference on Advanced Information Networking and Applications (AINA), pp. 27–33 (2010)
3. Abdelwahab, S., Hamdaoui, B., Guizani, M., Rayes, A.: Enabling smart cloud services through remote sensing: an internet of everything enabler. Internet of Things J. IEEE **1**(3), 276–288 (2014)
4. SNIA Cloud Storage Initiative. In: Implementing, Serving, and Using Cloud Storage. Whitepaper (2010)
5. Aazam, M., Khan, I., Alsaffar, A.A., Huh, E.-N.: Cloud of things: integrating 1 internet of things and cloud computing and the issues involved. In: Proceedings of 2014 11th International Bhurban Conference on Applied Sciences and Technology (IBCAST) Islamabad, 14th-18th January, 2014 (2014)
6. Harnik, D., Pinkas, B., Shulman-Peleg, A.: Side channels in cloud services: deduplication in cloud storage. Secur. Privacy IEEE **8**, 40–47 (2010)
7. Guo-Zi, S., Yu, D., Dan-Wei, C., Jie, W.: Data backup and recovery based on data de-duplication. In: 2010 International Conference on Artificial Intelligence and Computational Intelligence (AICI), pp. 379–382 (2010)
8. Kumar Bose, S., Brock, S., Skeoch, R., Shaikh, N., Rao, S.: Optimizing live migration of virtual machines across wide area networks using integrated replication and scheduling. In: 2011 IEEE International Systems Conference (SysCon), pp. 97–102 (2011)
9. Bose, S.K., Brock, S., Skeoch, R., Rao, S.: CloudSpider: combining replication with scheduling for optimizing live migration of virtual machines across wide area networks. In 2011 11th IEEE/ACM International Symposium on Cluster, Cloud and Grid Computing (CCGrid), pp. 13–22 (2011)
10. Mandagere, N., Zhou. P., Smith, M.A., Uttamchandani, S.: Demystifying data deduplication. In: Proceedings of the ACM/IFIP/USENIX Middleware '08 Conference Companion, Leuven, Belgium (2008)
11. Bhagwat, D., Pollack, K., Long, D.D.E., Schwarz, T., Miller, E.L., Paris, J.F.: Providing high reliability in a minimum redundancy archival Sto rage system. In: 14th IEEE International Symposium on Modeling, Analysis, and Simulation of Computer and Tele Communication Systems, 2006. MASCOTS 2006, pp. 413–421 (2006)

12. Hartman, R.D.: Architecture and measured characteristics of a cloud based internet of things. In: 2012 International Conference on Collaboration Technologies and Systems (CTS), IEEE, pp. 6–12 (2012)
13. Yinjin, F., Hong, J., Nong. X., Lei. T., Fang. L.: AA-Dedupe: an application-aware source deduplication approach for cloud backup services in the personal computing environment. In: 2011 IEEE International Conference on Cluster Computing (CLUSTER), pp. 112–120 (2011)
14. Lei, X., Jian, H., Mkandawire, S., Hong. J.: SHHC: a scalable hybrid hash cluster for cloud backup services in data centers. In: 2011 31st International Conference on Distributed Computing Systems Workshops (ICDCSW), pp. 61–65 (2011)
15. Yang, Z., Yongwei, W., Guangwen, Y.: Droplet: a distributed solution of data deduplication. In: 2012 ACM/IEEE 13th International Conference on Grid Computing (GRID), pp. 114–121 (2012)

Printed in the United States
by Baker & Taylor Publisher Services